Menstruation

Menstruation

A Cultural History

Edited by

Andrew Shail

and

Gillian Howie

First published in 2005 by
PALGRAVE MACMILLAN
Houndmills, Basingstoke, Hampshire RG21 6XS and
175 Fifth Avenue, New York, N.Y. 10010
Companies and representatives throughout the world.

PALGRAVE MACMILLAN is the global academic imprint of the Palgrave
Macmillan division of St. Martin's Press, LLC and of Palgrave Macmillan Ltd.
Macmillan® is a registered trademark in the United States, United Kingdom
and other countries. Palgrave is a registered trademark in the European
Union and other countries.

ISBN-13: 978–1–4039–3935–7 hardback
ISBN-10: 1–4039–3935–7   hardback

This book is printed on paper suitable for recycling and made from fully
managed and sustained forest sources.

A catalogue record for this book is available from the British Library.

Library of Congress Cataloging-in-Publication Data

Menstruation : a cultural history / edited by Andrew Shail and
Gillian Howie.
p. cm.
Includes bibliographical references and index.
ISBN 1–4039–3935–7 (cloth)
1. Menstruation – History. 2. Menstruation – Folklore – Cross-cultural
studies. I. Shail, Andrew, 1978– II. Howie, Gillian.

QP263.M47 2005
612.6'62—dc22   2005043427

10 9 8 7 6 5 4 3 2 1
14 13 12 11 10 09 08 07 06 05

Contents

Notes on Contributors

Luigi Arata teaches at the scientific Lycée 'Arturo Issel' in Italy. He held a scholarship in Ancient Greek Literature at the University of Genoa and is now included in international projects such as Commentaria et Lexica Graeca in Papyris Reperta and Ancient Greek Grammars' Lexicon. He has published on the history of Greek science and medicine and on ancient Greek rhetoric, and has edited editions of Aeschylus' tragedy *Choephorae* (1999) and Lysias' oration *For the Invalid* (2001). Forthcoming work includes an edition of Sophocles' tragedy *Antigone*.

Ariane M. Balizet is a PhD candidate in English at the University of Minnesota, USA. Her dissertation investigates the metaphorical and material meanings of blood on the Renaissance stage. She has presented work on topics ranging from the poetry of George Herbert to feminist pedagogy in the Shakespeare classroom. Her article 'Teen Scenes: Recognizing Shakespeare in Teen Film' appears in *Almost Shakespeare: Reinventing his Works for Cinema and Television* (2004).

Bettina Bildhauer is Lecturer in German at the University of St Andrews, UK. She has worked extensively on menstruation, as well as blood and gender more generally, in medieval fictional, devotional and medical writings. Some of her work on this is published in her *Medieval Blood* (2005), as well as in an article on bloodsucking in *Consuming Narratives* (2002). She has also co-edited, with Robert Mills, *The Monstrous Middle Ages* (2003), and is currently working on a book on representations of the Middle Ages in German cinema.

Helen Blackman is a research fellow in medical history at the University of Exeter, UK. She completed her PhD, 'Women, savages, and other animals: The comparative physiology of reproduction, 1850–1914', at the University of Manchester in 2001. She is the author of articles on the history of embryology, zoology and reproductive physiology. Forthcoming work includes a monograph on the history of zoology and medical education in late-nineteenth century England.

Monica H. Green is Professor of History at Arizona State University, USA. She teaches courses in Medieval History, Women's History, and the History of Science and Medicine. She has published extensively on various aspects of the history of women's medicine, including the collections *Women's Healthcare in the Medieval West: Texts and Contexts* (2000) and *The 'Trotula': A Medieval Compendium of Women's Medicine* (2001). Forthcoming work includes *The Masculine Birth of Gynaecology* and a general study of the medical school of Salerno.

Gabrielle Hiltmann is Lecturer in Philosophy at Basel University, Switzerland. She is the co-editor of 'Les disciplines en jeu', an issue of the French-Swiss journal *Nouvelles Questions Féministes* (2004), and the author of a work on Ludwig Wittgenstein's method. She has published articles on French contemporary philosophy, phenomenology, hermeneutics, methodology, interdisciplinarity, ethics, and feminist philosophy. Forthcoming work includes the collection *Alterity and Sex/Gender – Phenomenological Reflections in Ethics*.

Gillian Howie is a Senior Lecturer in Philosophy at the University of Liverpool, UK, and Director of the Institute for Feminist Theory and Research. She is author of *Deleuze and Spinoza: Aura of Expressionism* (2002), editor of *Critical Quarterly*'s special issue on higher education (2005), and editor of *Women: A Cultural Review*'s special issue on gender and philosophy (2003). She is also co-editor of *Gender, Teaching and Research in Higher Education* (2001) and *Third Wave Feminism: A Critical Exploration* (2004), the first two texts of the IFTR conference series. She is currently working on feminism and (historical) materialism.

Dianne E. Jenett is core faculty in Humanities and Co-Director of the Women's Spirituality Master's Programme at New College of California in San Francisco and adjunct faculty in the Depth Psychology Master's Programme at Sonoma State University, USA. She has published articles on transpersonal research methodologies for the social sciences and women's festivals and ritual in South India. She received her PhD from the California Institute of Integral Studies.

Cathy McClive completed her PhD, 'Bleeding Flowers and Menstrual Moons: A History of Menstruation in France, c. 1495–1761' at Warwick University, UK in 2004. She is currently researching medical expertise in early modern France on a Leverhulme Trust postdoctoral award at the École Pratique des Hautes Études at the Sorbonne. She has published in English and French on pregnancy, menstruation and medical authority.

Zahra Meghani is a doctoral candidate in philosophy at Michigan State University, USA. She is co-editing a book on the medical ethicist Michael Ryan.

Marie Mulvey-Roberts is a Reader in Literary Studies at the School of English and Drama at the University of the West of England, UK. She is the author and editor of nearly forty books and databases including *Defining Gender, 1450–1910: The Mind and Body* (2003–07), *Gothic Fiction* (2001–03) and *Sex and Sexuality, 1640–1940: Literary, Medical and Sociological Perspectives* (2001). She has also published articles on gender, the female Gothic and the body, and literature and medicine and is the co-editor of the journal *Women's Writing*. Forthcoming work includes editions of the first novel and collected letters of Rosina Bulwer-Lytton.

Rebecca Munford is a Lecturer in Literature, Culture and Theory at the University of Manchester, UK. The co-editor of *Third Wave Feminism: A Critical Exploration* (2004), she has also published articles on third wave feminism, Angela Carter and the Gothic. Her current projects include a monograph on twentieth-century women writers and the Gothic, an edited collection on the politics of intertextuality in Angela Carter's writing and *(Un)popular Feminisms* (2006).

Haviva Ner-David is completing her doctorate, on Jewish menstrual ritual impurity, at Bar Ilan University, Israel. She is the author of *Life on the Fringes: A Feminist Journey Towards Traditional Rabbinic Ordination* (2000) and is studying towards Orthodox rabbinic ordination.

Hollis Robbins is Assistant Professor of English at Millsaps College, USA. She obtained a PhD in English from Princeton in 2003. She is co-editor of *In Search of Hannah Crafts: Essays on The Bondwoman's Narrative* (2003). Journal articles include 'The Emperor's New Critique' (*New Literary History*, 2003) and 'Flushing Away Sentiment: Water Politics in Edith Wharton's *The Custom of the Country*' (*Ellipses*, 2000). She is currently writing a book on Margaret Clephane Compton's epic poem, *Irene*.

Andrew Shail is Lecturer in Film at the University of Northumbria in Newcastle, UK. He has published on male feminism in *Third Wave Feminism: A Critical Exploration* (2004) and on the cinematic body in *The Matrix Trilogy: Cyberpunk Reloaded* (2005). He is currently working on articles on menstruation and consumer waste, the first Kotex advertisements, early UK picture personalities and film fan magazines, and a monograph on the origins of the modern super hero.

Michael Stolberg is Professor of History of Medicine at the Universität Würzburg, Germany. He has published various books on the history of public health and hygiene, most recently *Homo patiens. Krankheits- und Körpererfahrung in der Frühen Neuzeit* (2003), a study of early modern lay perceptions and experiences of the body and its diseases. His work on gender history and women's history includes articles on the origins of the eighteenth-century anti-masturbation campaign, the history of pre-menstrual complaints and menopause and a critique of Thomas Laqueur's influential claims on the anatomy of sexual difference in early modern medicine.

Julie-Marie Strange is Lecturer in Modern British History at the University of Manchester, UK. She has published on the cultures of death in late Victorian Britain in *Past and Present, Social History, Journal of Contemporary History* and *Mortality* and in her monograph *Death, Grief and Poverty, 1870–1914* (2005). She has published articles on menstruation in *Women's History Review* and *Social History of Medicine* and is currently working on a monograph exploring the dialectic between popular and medical understandings of menstruation at the turn of the twentieth century. She is co-editor of the forthcoming new edition of *Twentieth-Century Britain*.

Jacqueline Thomas is a doctoral candidate in English at the University of Texas at Austin, USA. Her study of menstruation in the works of James Joyce involves an anthropological re-examination of the archaic and classical tropes of femininity which Joyce employed.

Sharra L. Vostral is Assistant Professor in Science and Technology Studies at Rensselaer Polytechnic Institute, USA. She is completing a book about the history of menstrual hygiene technologies in the United States, and the ways these technologies have shaped understandings of menstruation as an organic process. Her research centres upon the history of technology in relation to gender and women's bodies, as well as the history of sexuality and medicine.

Sabine Wilms is Senior Editor at Paradigm Publications in Taos, New Mexico, USA. She is co-author of *Getting to Know Chinese Medicine* (2005), the translator of clinically-oriented texts on Chinese medicine and the author of articles on medieval Chinese gynaecology. Her research interests include the history of Chinese medicine and gynaecology, the transmission of Chinese medicine to the West, and the cross-cultural study of the female body, women's healthcare, childbirth, and menstruation.

Acknowledgements

Andrew would like to thank Stacy Gillis for feedback, support, advice and sustenance, and Rebecca Munford for always answering email queries. Thanks also to all at the Institute of Feminist Theory and Research, particularly Gill Howie, for putting their faith in him, and to Ashley Tauchert for help and encouragement at all points, including her guidance on the postgraduate work that started it all.

Both editors are grateful to Laura Perrett, Anna Hunt and Yiota Vassilopoulou for their excellent work on *Menstruation: Blood, Body, Brand*, the conference from which this collection traces its origins. We would also like to thank all at Special Collections at the University of Newcastle-upon-Tyne's Robinson Library, as well as all at Special Collections and at the Bill Douglas Centre, both at Exeter University Library.

List of Figures

The authors and publishers have made every effort
to trace holders of copyright in the figures. If any has
inadvertently been overlooked, the publishers will be
pleased to make the necessary arrangements at the
first opportunity.

Introduction

'Talking Your Body's Language': The Menstrual Materialisations of Sexed Ontology

Andrew Shail and Gillian Howie

> Is there a history of how the duality of sex was established, a genealogy that might expose the binary options as a variable construction?
>
> (Butler, Gender 7)

An advertisement for Tampax Compak, shown only in cinemas, shows a crowded cinema auditorium. A warning sign and voiceover follows, indicating that 'the following manoeuvre must only be undertaken with the proper protective gear'. A woman is then shown moving along a crowded row to her seat, the camera following her bottom and hips closely. She sits down and the safety gear is revealed to be the Tampax Compak tampon.

As with all commodities in industrial culture, menstrual products are the result of a complex process of production controlled by various material and social factors. The modern mass-production and circulation of 'feminine hygiene' goods has been facilitated by new manufacturing, distribution and broadcasting technologies, changes in the labour market and the adoption of the modern, 'democratic', concept of the market. To be effective as consumables, these 'hygiene' goods must be marketed in a way that resonates with personal and social values, meaning that their advertising presents us with a prism through which to see complex social patterns (Williams 321, 334). Shifts in the marketing of these products may reflect much wider changes in social and economic patterns. But tendencies in menstrual marketing also reflect, and influence, a more embedded genealogy. As the above advertisement indicates, the term 'feminine protection products' has recently begun to replace 'feminine hygiene products' in industry vocabulary. The most explicit and loudest form of discussion of the menses, advertising has gradually secured the tenacity of 'protection' as a set of ideas adhering to

1

the menstrual (and so the female) body, and has done so in spite of the apparent waning of the idea that menstruation constitutes a hygienic predicament.[1] Whereas a simultaneous industry-wide trend towards describing the disposable sanitary product as an accessory reflects the naturalisation of neo-colonial economic relations within a notion of 'girliness',[2] the discourse of 'protection' seems to relate to a lengthy history of thinking about femaleness which perceives undecidable limits and demarcates the female body as a danger to itself.

Women seem to exemplify the effect of, and to manifest the threat of, corporeal chaos (Shildrick 15). 'Can it be that in the west, in our time', Elizabeth Grosz asks, 'the female body has been constructed not only as a lack or absence but with more complexity, as a leaking, uncontrollable, seeping liquid; as formless flow; as viscosity, entrapping, secreting; as lacking not so much or simply the phallus but self-containment'; so not just as leaky but as formless? Women's corporeality, she writes, 'is inscribed as a mode of seepage' (*Volatile* 203). Ideas surrounding menstruation could therefore be expected to be a major site of this inscription, with promises of 'protection' implicitly promising compensation for a female body presented as an unruly threat to the idea of bodily completeness. This inscription is continuous with the marking-out of maleness as the model of the complete, sex-neutral human body. A certain (perhaps intentional) ambiguity, however, exists in Grosz's question. Menstruation might be one of the ways in which the female body exceeds or resists representation, hence its widespread demonisation within modern Western culture and the need for an offer of 'protection'. But the idea of the female body as threatening to notions of bodily completeness because of unique flows, as less representable and less conscribed than the male body, as a disruptive outside to the body as conscribed within systems of representation, follows from purely historical concepts of sexed bodies as entities that present as distinct. The female body's 'undecidable limits' are, rather, the result of its *representation* as less representable, this representation having acted to disguise the fact that the body marked female is one of the *most* represented cultural items.

This collection excavates past menstrual bodies and the contexts in which they were created. What is clear from the chapters here is that menstruation understood through the prerogatives of modern mass consumption is only the latest in a series of dominant understandings of menstruality and the body marked female. Although menstruation has not always been primary in the sexing of the body, it has been closely tied into valuations of the body. As Moira Gatens points out, in Western thought, '[t]he most superior minds suffer least from the intrusions of the body. Women are most often understood to be less able to control the passions of the body and this failure is often located in the a priori *disorder* or anarchy of the female body itself' (50; emphasis in original). One strand of a history of menstruation is the examination of how ideas about specific mechanisms have served to contain certain bodies. As Linda Alcoff summarises, man positions woman to be

> defined, delineated, captured – understood, explained and diagnosed – to a level of determination never accorded to man himself. ... Despite the variety of ways in which man has construed her essential characteristics, she is always the Object, a conglomeration of attributes to be predicted and controlled along with other natural phenomena. (434–5)

Menstrual discourse has been one of the ways in which the female body has been more intensely described, and the female-embodied described as more bodily, than maleness. Menstruation is a focal point for a logic of female bodily peculiarity that, representative or not, serves to provisionalise the inclusion, as human, of the female-embodied through refusing to see the male body in the same terms, in spite of the fact that its homologies with the 'opposite' sex are numerous. As Iris Marion Young points out, 'men's bodies are at least as fluid as women's' (193). Understanding menstruation as evidence of female-embodiment's superfluity of bodily flows systematically elides this homology. Widespread notions of a totally sexed body are also achieved by one of the major impacts of much menstrual language, the commonplace contemporary description of women alone as 'thinking with their glands' reducing the body marked female to a function of the uterus and ovaries (de Beauvoir 15). For Elizabeth Grosz, as for several other feminist contemporaries,

> The reinscription of sexual morphology in terms more conducive to women's corporeal and sexual anatomy, beyond the problematic of lack, would entail a thoroughgoing transformation of the social meanings of sexual difference, and consequently the constitution of different body images for the two sexes. (Grosz, *Volatile* 82)

One impulse behind this collection, therefore, is the excavation of alternative modes of embodiment that may be crucial to a project that feminism has made clear is necessary.

The menstrual body is a body of, in one respect, certainty. The groin signifies sex difference categorically in the menstrual blood and shed mucous membrane of the female-embodied. For example, some opera singers' contracts require them not to perform on the first and second days of their period, because of alterations in vocal abilities referred to as 'menstrual cycle dysphonia'. High nocturnal body temperature is commonplace during the peri-menstruum. These and many other understandings are foremost amongst the forces that might lead one to question whether it could be said of certain areas of the body that '[t]here is nothing inherent in these regions and zones that makes them more suitable for culturally representing sexual difference – many others would have served this function just as well; what culturally marks sexual difference is biologically arbitrary, conventional' (Grosz, *Volatile* 203). Concessions of pre-discursive sex difference are perhaps most frequently made with reference to the menses. But what is menstruation? Even setting aside alterations in the incidence of menstruation illustrated by Joan Jacobs Brumberg's oft-cited article on lowering age of menarche in the nineteenth-century West, throughout history the referent of this term (and its equivalents) has varied constantly between the extremes of mere blood flow and a total body phenomenon, as can be seen in the fact that the term 'the period' presently connotes just blood flow for many, but for just as many others connotes a 24-day set of events. The phenomenon has been understood to occur without blood flow, and vaginal blood flow has also gone un-linked with any wider bodily process.

Discussing, in Grosz's words, 'what culturally marks sexual difference' with a focus on menstruation involves historicising not just the ways in which sex difference is interpreted, but also regard for sex as a matter of difference, and even the notion that the body's primary attribute is its 'fact' of being sexed. Elementary understandings of sex-as-difference and the-body-as-sexed both draw on the understanding of menstruation as a process of very specific bodies. This is not to argue that men should be seen as having menstruated too, but that menstruation is one of the ways that corporeality has been rendered intelligible as a (sexed) body.[3] The title of this introduction, the tag-line of contemporary *Always* television advertising, in addition to being deliberately ironic, reflects an attempt to re-conceive the history of menstruation in this light. The capacity to speak the language of a body is one of the ways in which authority over that body has been achieved and that body's status decided. But, as the chapters in this collection point out, what we discover when we discuss the body is, mostly, a language. Signifying practices have not simply produced ideas of a neutral body. They have actually rendered corporeality intelligible and discoverable as 'the body'. They have 'intextuated' corporeality, to use Grosz's term (*Space* 35), 'materialised' it, to use Judith Butler's, through regulatory norms (*Bodies* 4–12), and menstruation ranks first on the list of sexually differential biological processes that, Grosz explains, 'must be signified in all cultures' as part of this process of intextuation (*Space* 36). The title of this introduction is also intended to imply that texts prefaced by claims to 'talk your body's language' enact a demarcation, a marking-out, of what constitutes the body. Lastly, although the texts discussed in this collection, like the text of Tampax Compaq advertising, relate in varying degrees to grand social patterns, the texts produced by these patterns are one of the mechanisms that render corporeality intelligible as 'the body'. Sexed, patterned in its movements, with certain easily-discarded parts and a certain level of resilience, incapable of certain acts, the body is produced by every description of it.

The chapters in this collection avoid understanding menstrual taboo solely as a reaction to the activities of the body. This follows from the recent insistence that 'the body must be reconceived, not in opposition to culture but as its preeminent object' (Grosz, *Space* 32). One key aspect of recent developments in feminist theory is this conviction that 'the sexed body can no longer be conceived as the unproblematic and factual base upon which gender is inscribed, but must itself be recognized as constructed by discourses and practices that take the body both as their target and as their vehicle of expression' (Gatens 70). Citing the body as opposed to culture privileges it, and menstruation has been constituted, through precisely this citing, as a site of knowledge, divulging truths. Penelope Shuttle and Peter Redgrove's *The Wise Wound* (1978) referred to menstruation as a truth that the lies of culture cannot resist. '*Suppose that society is a lie, and the period is a moment of truth which will not sustain lies*' (58; emphasis in original), they invited, so invoking menstruation as a truth about which speech must be invigorated by referring to 'a conspiracy of silence' (13). One contemporary regime of knowledge centred on menstruation is the citing of its onset, the competition amongst teenagers in the West about 'getting it', often obsessively pursued. Of course, discourse about who 'has it' operates socially, forcing, in this case, classed good health and capacity for reproduction to masquerade as knowledges. Menstruation's discursive existence is not an inscription of sexed materiality on the neutral body but a rendering intelligible of the body solely in terms of sexed materiality. If menstruation is implicitly the expression of a sex-gender arrangement and explicitly something unclean about which a shamed silence must be retained, then it is sexual difference (sex-as-difference) about which we cannot speak. Where changes occur, therefore, menstruation will be found to be a site of sex-based conflict. As a component of the historical female subject, even those components of menstrual narrative that can be loosely defined as limiting, discriminatory, laws are generative laws nevertheless, as much forming as limiting the tabooed 'woman'.[4]

It is important to acknowledge the usefulness of holding up menstruation as one dimension of bodily difference which discourse on sex has tried to turn into an aspect of women's 'castration', a position of secondariness to men. Regardless of the varied ways in which menstruation depicts the female body, dominant representations serve to ground the devaluation of women by men (Shildrick 14). The cultural history of menstruation might be a history of a succession of ways for denying difference. However, it might also be a history of the ways in which difference is established as pre-social, where invoking menstruation as a *hidden* difference approaches the activities of the *sexed* body as the most important of pre-cultural facts. Each citation of the menstruality of the female body, each use of menstruation to refer to 'fifty percent of humanity', materialises the body as a sexed entity, so bringing sex to the fore as a salient factor of the body's being, rendering bodies unambiguous according to a single criterion. Menstruation is a case in which concessions of pre-discursive sex difference are often made. However, '[t]o "concede" the undeniability of "sex" or its "materiality" is always to concede some version of "sex", some formation of "materiality" ' (Butler, *Bodies* 10). What was 'the female body' at the historical configuration when – as Sabine Wilms's chapter shows – less than menstruation was considered to be menstruation? And what was 'the female body' at the historical configuration when – as Cathy McClive's chapter shows – more than menstruation was considered to be menstruation? Declaring the aggregate similarity of bodies so unstably distinct might itself be another gesture of seeing 'women' as castrated men, but 'state of nature' stories of sex difference, like sex-distinction along the lines of whether a body can give birth, disguise, and allow us to take for granted, 'the intricate and pervasive ways in which patriarchal culture has made that difference its insignia' (Gatens 73). Sex-aligned difference is the condition for particular social arrangements.

Menstruation is one of the primary ways in which the body has been rendered intelligible as sexed, appearing to denote most explicitly the biological basis of stable dimorphic sex-based categorisation. Menstruation discourse both foists upon the female-embodied a 'body imaginary'[5] which originates outside the purview of their own sexed embodiment, and continues to materialise the body as sexed through micro-engineering female morphology. It both furthers the alienation central to female-embodiment *and* works to maintain the internal stability of the terms 'men' and 'women'. Menstruation is an extremely common conversation subject for Western women when asked or motivated to discuss their femaleness, *and* the secret consecrated as a site of truth by, for example, the special sex education classes from which young men tend to be excluded. Gender is not false consciousness but stylisation of the body. Moira Gatens notes that 'in our culture it is associated with shame and modesty – both characteristically feminine attributes' (9–10). The language of menstruation has helped to ensure sexual difference, doing so both through materialising the sexed body – deciding whether menstruation is an event, object, substance or process – and through influencing the event, object, substance or process. Menstrual language and the phenomenon of menstruation are, after all, only distinct insofar as psychology and physiology are distinct.

This collection attends throughout to the 'unacknowledged philosophical underpinnings of dominant representations of sexual difference' (Gatens vii), and to how it is that many different events we retrospectively and often over-confidently call 'menstruation' have been the subject of accounts, a language for talking about the body which constitutes, at any one time, a topic constructed by the statements used to refer to it. For example, a January 2003 Channel 4 documentary, *Bloody Women*, discussed menstruation as follows: 'Abnormalities in brain chemicals the week before menstruation mean that some women become a bit obsessive in their behaviour.' Showing teenagers buying large bars of chocolate, the programme ascribed a certain behaviour involving food not to a virulent self-loathing inculcated in the female-embodied from the time they are encouraged to call themselves women but to a sexed constitution. Simultaneously, the statement made by showing the teenagers buying chocolate is one of the 'discursively constrained performative acts that produce the body through and within the categories of sex' (Butler, *Gender* viii). While remaining sensitive to the possibility that the guiding impulse may be a-historical, an urge to ignore a history of sexed oppression, in what hopes to contribute to Foucault's intended 'labour of diverse enquiries' ('What is Enlightenment' 50) we are asking 'how the criteria of intelligible sex operates to constitute a field of bodies, and how precisely we might understand specific criteria to produce the bodies that they regulate' (Butler, *Bodies* 55). This collection focuses on signifying practices that both represent the menstruating body and constitute it as a site of knowledge. While the female body has been abused by the social structures which interpret menstruation for us, it is simultaneously constituted by them too. Partially for this reason, but also because menstrual narrative is not reducible to oppression, as well as because a cultural history of the language of anything is a historical story of more than just language, the aim of this collection is also to demonstrate when and how menstrual narrative has not been reduced to functional iteration of misogynistic values.

The following chapters seek to establish both theoretical and chronological frameworks for the study of the language of menstruation, and include an introduction to some of the major bibliographical and archival sources for the study of this history. Bearing in mind that this includes two (albeit often interrelated) streams – the *construction* and the *representation* of menstruation – it is split into two sections. The first, Science and Medicine, traces the construction of menstruation as an object of knowledge from the ancient Greek fifth to third century BCE *corpus Hippocraticum* to the early 21st-century practice of menstrual suppression. The second, Myth and Culture, examines signifying practices and belief systems that have determined and invested in images of menstruation and which have preserved menstrual discourse in coded forms. The chapters in this section deal with how menstruation is represented, with the documentary evidence these representations have preserved about the changing materialisation of sex and gender, and with how these representations have guided the materialisation of sex and gender. The categories of 'science and medicine' and 'myth and culture' were, from the outset of this collection, never intended to suggest that these categories have ever operated independently. Indeed, the very title of this collection – *Menstruation: A Cultural History* – draws upon the Foucauldian model of discourse as a cultural event. As representations of medical configurations of menstruation are not simply windows on discourse but shape its development, strands of 'empirical' thought on menstruation are registered in several of the chapters on myth and culture. The two sections can be easily super-imposed and should be understood as parts of a single history.

Subjects constantly discover parts of their bodies through socially shared significances, and these significances have been, in part, transmitted through and encoded in a variety of media, ranging, within just the last 200 years, from the blood-initiated sleep of Briar Rose (better known as Sleeping Beauty), through the menarcheal insanity of William Peter Blatty's novel *The Exorcist* (1971) and William Friedkin's film version in 1973 (Shuttle and Redgrove 240–55), the pubertal leg-separation blood of *The Little Mermaid* (1836), the change from silver to ruby slippers between L. Frank Baum's novel *The Wonderful Wizard of Oz* (1900) and Victor Fleming's film *The Wizard of Oz* (1939), to the economy of secretions in Ivan Reitman's *Ghostbusters II* (1989). But where cultural texts are the object of study, the conclusion that a text is 'about' menstruation is insufficient – the real question is 'what record does this constitute about the history of the sexed body?' What is the encoded object? The chapters in both sections recognise that menstruation is at the centre of vast social and economic arrangements, often being influenced by circumstances that ostensibly have little to do with it. In addition, because little about menstruation is implicit in the 'process' (our understanding of menstruation as a 'process' being itself historically contingent, as many of these chapters show), it is achieved semantically each time that it is referred to. This incredibly malleable construction of menstruation – varying even in the last three centuries from a substance to a secretion to a cycle to a pathology – means not only that menstruation is constructed in arenas beyond those medical discourses which construct it as an object of knowledge, but that many more signifying acts concern menstruation than may be immediately apparent. This is not to re-evoke menstruation as a 'hidden' secret, and therefore as an omnipresent truth to be uncovered behind all texts, but to recognise how many discourses constitute segments of the most variable of 'a-histories', that of femaleness and maleness.

Notes

1. See, for example, 'Mother' par. 1.
2. This is also part of a widespread appropriation of the language and symbols of 'girlie culture' (Munford 145).
3. The relatively high proportion of chapters in this collection that take male menstruation as their subject reflect the impulse to investigate menstruation from this point of view.
4. On power as generative see Foucault (*The Will to Knowledge* 36–49).
5. For an erudite employment of this terms see Brain (158).

Works cited

Alcoff, Linda. 'Cultural Feminism Versus Post-Structuralism: The Identity Crisis in Feminist Theory'. *Feminism and Philosophy: Essential Readings in Theory, Reinterpretation, and Application*. Ed. Nancy Tuana and Rosemarie Tong. Boulder: Westview, 1995. 434–56.
Anderson, Hans Christian. 'The Little Mermaid'. 1836. *Fairy Tales*. 4 Vols. Trans. R. P. Keigwin. Odense, Den.: Skandinavisk Bogforlag and Flensted, 1976. Vol. 1. 122–70.

Baum, L. Frank. *The Wonderful Wizard of Oz*. 1900. Harmondsworth: Penguin, 1995.
de Beauvoir, Simone. *The Second Sex*. 1949. London: Vintage, 1997.
Blatty, William Peter. *The Exorcist*. London: Blond and Briggs, 1971.
Bloody Women. Prod. Philippa Ross. RDF Media. Channel 4. 8 Jan. 2003.
Brain, Josephine. 'Unsettling "Body Image": Anorexic Body Narratives and the Materialization of the "Body Imaginary" '. *Feminist Theory* 3.2 (Aug. 2002): 151–68.
Brumberg, Joan Jacobs. ' "Something Happens to Girls": Menarche and the Emergence of the Modern American Hygienic Imperative'. *Journal of the History of Sexuality* 4.1 (July 1993): 99–127.
Butler, Judith. *Bodies That Matter: On The Discursive Limits of 'Sex'*. London: Routledge, 1993.
——. *Gender Trouble: Feminism and the Subversion of Identity*. London: Routledge, 1990.
The Exorcist. Dir. William Friedkin. Perf. Ellen Burstyn. Warner Bros, 1973.
Foucault, Michel. 'What is Enlightenment?' 1978. *The Foucault Reader*. Ed. Paul Rabinow. London: Penguin, 1984. 3–50.
——. *The Will to Knowledge*. 1976. Vol. 1 of *The History of Sexuality*. 3 vols. 1976–1984. Trans Robert Hurley. London: Penguin, 1981.
Gatens, Moira. *Imaginary Bodies: Ethics, Power and Corporeality*. London: Routledge, 1996.
Ghostbusters II. Dir Ivan Reitman. Perf. Bill Murray. Columbia Pictures, 1989.
Grosz, Elizabeth. *Space, Time, and Perversion: Essays on the Politics of Bodies*. London: Routledge, 1995.
——. *Volatile Bodies: Toward a Corporeal Feminism*. Bloomington: Indiana University Press, 1994.
'Mother to Daughter. Choosing Feminine Protection Products'. *The One. The Only. Tampax*. 2002. Tampax. 14 Nov. 2004. <http://www.tampax.com/en_us/pages/m2d_main.shtml?pageid = ar0080>.
Munford, Rebecca. 'Wake up and Smell the Lipgloss: Gender, Generation and the (A)politics of Girl Power'. *Third Wave Feminism: A Critical Exploration*. Ed. Stacy Gillis, Gillian Howie and Rebecca Munford. Basingstoke: Palgrave, 2004. 142–53.
Shildrick, Margrit. *Leaky Bodies and Boundaries: Feminism, Postmodernism and (Bio)Ethics*. London: Routledge, 1997.
Shuttle, Penelope and Peter Redgrove. *The Wise Wound: Menstruation and Everywoman*. 1978. Harmondsworth: Penguin, 1980.
Williams, Raymond. 'Advertising: The Magic System.' *The Cultural Studies Reader*. Ed. Simon During. London: Routledge, 1993. 320–36.
The Wizard of Oz. Dir. Victor Fleming. Perf. Judy Garland. Metro-Goldwyn-Mayer, 1939.
Young, Iris Marion. *Throwing Like a Girl and Other Essays in Feminist Philosophy and Social Theory*. Bloomington: Indiana University Press, 1990.

Part I
Science and Medicine

1
Menses in the *corpus Hippocraticum*

Luigi Arata

'Gynaecology' in the *corpus Hippocraticum*

The most celebrated physician in ancient Greece was Hippocrates
(c. 460–380 BCE). Ascribed to him are roughly eighty anonymous medical
works that became part of the collection of the Library of Alexandria
after c. 200 BCE. These diverse writings, all in some way linked with
Hippocrates, are now known as the *corpus Hippocraticum* although it contains many treatises that actually have little in common with Hippocrates'
school and a few not written during his lifetime. This variation undermines the common assumption that in the *corpus* only one 'Hippocratic'
medicine or one medical school is represented. The same mistake –
understanding the *corpus* as a unity, and consequently that all Greek
medicine represented therein is in accord – has had the effect of confusing scientific discussions about the specific characteristics of what
ancient tradition defined as 'Hippocratic', and, therefore, muddied discussions concerning the aspects of the treatises that we would now call
gynaecological. Added to this is the erroneous notion that ancient Greek
science was constructed purely on philosophical deductions rather than
direct observation. In this chapter, I shall establish that observation was
in fact central to medical practice and, in particular, to 'Hippocratic'
gynaecology, as well as pointing up the neutral objectivity with which
many of these medical practitioners regarded menses.[1] It is well known
that in the *corpus Hippocraticum* a series of works cover issues related to
delivery and to female physiology and that within this series there is a
gynaecological treatise, 'Feminine Diseases'. The latter is a vast work
which, according to the analysis of Hermann Grensemann, includes
three different treatises by three different physicians as contradictions,
differences in method and, not least, differences in the technical lexicon

strongly suggest. In order to clarify that the three approaches are distinct, I shall identify some of the differences between the three physicians, traditionally identified by the letters A, B and C.

Physician A, the earliest of the three, limited himself to describing the symptomatology of diseases and to giving therapeutic indications, amongst which the 'milkcure' is predominant (63. 128, 14). One of the main characteristics of his method is that the cure begins either by purifying the patient from an excess of two humours – bile and phlegm (37. 90, 2–3) – or with a lenitive intervention, which means confronting the most difficult, or inconvenient, symptoms (125. 268, 12–15). The physician's care towards female patients illustrates a new development in therapeutic style in that physician A advised that women, when weak or recovering, should *not* be subjected to violent therapies (52. 112, 2–3). Physician B was even more concerned than A with the needs of the female patient and is also sensitive to her psycho-physical condition (35. 82, 18; 43. 100, 22–102, 1). When he listed the symptoms of the diseases, he suggested that his readers should first establish the condition of the woman's body and only later that of her reproductive apparatus because the point of the therapy was to purify the whole body of the patient and thereby cure her uterus. In addition, he supported a therapy in which every symptom is opposed by an analogous and contrary intervention; for example, if a woman was afflicted with fever, she must be cooled. Physician B referred to medical tools to which A did not have access, such as the cupping-glass (71. 150, 20) and the infusor (66. 138, 11). However, because inner mechanisms causing diseases were largely unknown, his therapy concentrated on reducing symptoms. Humoural theory is important to B, but he extends pathogenic humours to include phlegm, bile and saltiness.

Physician C is the most recent of the three and the most explicit about his own work, to which he constantly referred. In addition to the authorship of the most advanced stratum in 'Feminine Diseases', authorship of 'On Generation, On the Nature of the Child and On Illnesses 4' can also be attributed to him. Conscious of having founded a new science in the treatment of 'disorders' specific to the uterus, he demonstrated a depth of thought that can be recognised in few other Hippocratic scientists. His account of the induction of abortion demonstrates a great breadth of experience ('On Generation' 13). He posited four humours – phlegm, bile, blood and water – in the human body. These humours are not pathogenic in themselves, although their imbalance is. Physician C's attempts to reconstruct feminine physiology can be understood as the apex of a medical tradition which, while it had its roots in popular belief, also referenced the latest theories of medical science. C proposed a classification of female diseases based on the quantitative and qualitative characteristics of menstrual flow. The first part of his gynaecological treatise concerns cases in which menses do not appear and in which blood flows more scarcely or more abundantly than normal followed by a consideration of cases in which the flow is full of phlegm, bile or water. The second part considers the same variations in cases of lochial (postpartum) purgation. In C's work, therefore, the flow becomes so fundamental as to constitute an 'instrument of organisation' by which his arguments are arranged.

Menses in C's work: analysis of 'Feminine Diseases' chapter 6

Of the three physicians who authored 'Feminine Diseases', only C considers the 'problem' of menarche. He understood menarche to occur when the veins supplying the genital organ become large enough to produce sufficient blood flow, similar to the way in which only male children are of certain age possess a sperm count ('On Generation' 2, 20). C was also the only one to provide a coherent description of menses as a natural phenomenon, explaining that his indications of normative amount and quality of the flow did not have absolute value as each woman should be studied in relation to her own conformation. C asserted that 'the blood of menses flows very dense and abundant in the middle days, while at the beginning and at the end they are scarcer and lighter' ('Feminine Diseases' ch. 6).[2] This corresponds with the modern model in which the menstrual cycle begins with light flow, increases and then decreases. He also claimed that 'in every woman, if she is healthy, the normal menstrual flow is of two Attic kotylae a little more or a little less abundant, and that in two or three days. A more or less long period marks a state of disease and of infertility' ('Feminine Diseases' ch. 6). One Attic kotyla is about 0.27 litres, meaning that, for C, the normal amount of menstrual blood would be more than half a litre. The modern medical community has unsuccessfully concerned itself with finding a reliable explanation of this datum (Dean-Jones, *Bodies* 89–92; Hanson, 'Signs' 11–12). This may be because it is difficult to estimate exactly how much menstrual blood is produced. Moreover, amounts vary from woman to woman and even from one menses to another. C, presumably trusting his female patients' evidence, exaggerated amounts that they had themselves exaggerated; dramatising blood loss is not uncommon.

As to the duration of the flow, the physician does *not* say (as nearly all commentators assume) that it must normally last 'two or three days', which would be in clear contradiction with his previous assertion that there were, at least two, middle *days*, in addition to both 'the beginning and ... the end' in which the flow is 'scarcer and lighter' suggesting, at least four days in total. The 'two or three days' referred to are evidently the days in which the flow is most abundant. At the end of the chapter he describes cases of a woman whose 'purgation' lasts for more than four days and another for whom it lasts for less than three days and considers both of them extreme cases ('On Air' 21; 'Prorrheticon 2' 28). This datum was widely agreed upon at the time. B relates the case of a woman who had not succeeded in becoming pregnant for a long time even though her menstrual cycle appeared quite regular, and advises her to apply a pessary (Goltz 226–8; McLaren 27–9) to her uterus for three days 'when it is the third or fourth day' of the flow ('Feminine Diseases' ch. 19).[3] In addition, physician A, in a passage describing a woman whose lochia has not occurred, adds, after the lochial flow has been achieved, the order of using a pessary 'when menses come, the first three days' (ch. 37).[4] The belief that the flow lasted more than three or four days was widespread, and was also included, for example, by the author of 'On the Eight-Month Foetus'.[5]

In this context, it is strange that data about cycle duration does not appear in chapter 6 of 'Feminine Diseases'. However, its monthly frequency can be located in the Greek word for menstruation, as the terms *epimēnia* and *katamēnia* derive from *mēn* 'the month'. C, unlike A and B, explicitly discusses the monthly aspect of menstruation in 'Feminine Diseases' (ch. 25) and 'On Generation' (ch. 15). There was widespread agreement among the physicians of the *corpus* on this aspect of menstruation (Dean-Jones, *Bodies* 94–101; cf. 'On Regimen' 1. 34; 'On the Eight-Month Foetus' 1; 'Prorrheticon' 2. 24). For example, the author of 'On the Seven-Month Foetus' states that 'menses appear in healthy women every month, because the month has a power of its own in the bodies' (9. 448, 5–7). Physician A proposed as anomalous the condition of a woman afflicted with an overabundance of phlegm: menses appear in her 'three times a month' ('Feminine Diseases' ch. 57; cf. Dean-Jones, *Bodies* 94–101).[6] In the same way, physician B observed, in a case of unnatural 'opening' of the uterus, that 'menses flow more viscous, abundant and frequent and the seed does not remain within' (ch. 166)[7] (cf. Dean-Jones, 'Bleeding' 185–91). C's innovation was that he supplied an explanation for the monthly occurrence of the menstrual cycle, basing it, on one hand, on the woman's moist nature and, on the other, on the monthly variations in body temperature.[8] Changes in temperature could, claimed C, affect the woman's humours, an effect to which she is more sensitive because of her higher moistness.

Following on from this, C instructs that

> You must draw the deductions by watching the woman's body and ask questions by comparing the menses with the previous ones, whether they are signs of a disease or not. If in fact they last fewer or more days than usual or flow less or more abundantly, they are signs of a disease, unless the woman's constitution is pathological and sterile. If instead it is so and [the woman's constitution] changes and becomes healthier, it is better. ('Feminine Diseases' 1)

In estimating physiological phenomena C made use of *natura procedit per gradus*, the principle that unexpected changes should be noted and, in this context, the hypothesis that changes are understood as the result of a pathogenic factor must be discarded. The implication is that none of these medium characteristics of the flow ought to be considered as a general norm, because each female patient, according to her physical and humoural constitution, will have a flow with its own characteristics. Therefore, C's claim is that any evaluation of menstruation must be based on flexible criteria. At the end of this chapter, C affirmed that 'menses flow like blood from a sacrificial victim and they are immediately coagulated, if the woman is healthy' ('Feminine Diseases' ch. 6; cf. 'Feminine Diseases' ch. 72, 'On Generation' ch. 18). In fact, in antiquity it was often, and inaccurately, assumed that an important characteristic of menstrual blood was that it coagulates quickly (Dean-Jones, *Bodies* 101–3; King, 'Blood' *passim*). A number of scholars have argued that this assumption is based on a misogynistic analogy between menstruating women and sacrificial animals. It could be argued, however, that physician C employed vivid similes to provide the greatest evidence for some of his observations, one of his peculiarities (Wenskus 393–406). Elsewhere in the *corpus*, instead of this simile, it is simply stated that menses have 'a beautiful colour' (*euchroa*) when the woman is healthy ('Prorrheticon 2' 24) and, vice versa, when the woman needs to be cured, her menses are 'colourless' (*achroa*) ('Aphorisms' 5. 36). The analogy of the sacrificial animal seems designed to specify that expelled blood, if healthy, should have a nice, bright colour (Wenskus 397), whereas, as stated in other chapters attributable to C, if unhealthy, it would look membranous 'like spider webs' ('Feminine Diseases' 9. 38, 8; cf. 29. 72, 11) or 'as if someone washed bloody material' (10. 124, 6).

C's relationship with his female patients

As I have shown, statements on menstrual phenomena were as scientific as any other medical claim at the time. Discrepancies between modern descriptions of menstruation and those in the *corpus* can be explained not by the physicians' deductive method or poor observation, but by the unreliability of their instruments of measure. Errors in estimating the amount of blood produced suggest that the physician would trust the affirmations of the women he treated. Indeed, C argued the necessity of a close relationship with his female patients, a relationship of confidence and communication which extended beyond the discomfiture of vaginal examination, with patients often being asked to practise self-examination outside contact time with the physician ('Epidemics' 5. 25. 224, 10–1; 'Feminine Diseases' 40. 96, 16ff.; cf. Hanson, 'Phaenarete' 162–78). Indeed, this would be one good explanation for the 'modern' description of menstruation produced by A, B and C. The assertion that Greek physicians did not maintain a dialogue with female patients certainly requires interrogation. Scholars have also for some time concentrated on the fact that the Hippocratic gynaecologists called menstruation *katharsis* 'purification', taking this to demonstrate that the 'normal' functions of the female body were considered to be pathological and therefore something which required treatment.[9] But this involves a misinterpretation of the medical term. The three gynaecologists observed simply that menstrual blood must be evacuated, otherwise there is a risk of the woman's physiology becoming blocked. Physician C dedicates one of the longest chapters in the treatise to the lack of menstrual flow. The suppression of menstruation was considered to be a common symptom not only of genital diseases, but also of more common ones, associated with nervous conditions and malnutrition. That women in ancient times could miss their normal menstrual cycle was explicable; their diet was often poor and they were often pregnant. The physician of 'Epidemics 5' tells the story of a woman who 'withheld menses for four years, except for a little bit. ... This woman became pregnant again and again' (11). In all these cases, whether menses had not been produced, or had been produced but not evacuated, the event was significant and, in some, dangerous: the *katharsis* must take place if the woman did not want to die. The importance attributed by Hippocratic physicians to purification was the result not of the association of femaleness with pollution[10] but demonstrative of a concern about the irregularity of regular phenomena and how this irregularity might indicate an underlying problem with the mechanisms of the organism.[11]

Moreover, according to physician C, during the puerperal period menses are the basis of the formation of the embryo's food ('On Generation' 18) and then of the formation of milk in the breasts ('Feminine Diseases' 73). C stated that the mother's blood becomes flesh or tissue by coagulation ('On Generation' 14–16). Further on (534, 10–536, 8), the timing of delivery is linked to the fact that the nourishment which comes from the mother's blood – after a period which is fixed at ten months ('On the Eight-Month Foetus' 4. 460, 4–7; 'On the Seven-Month Foetus' 4. 442, 14–15; cf. Diepgen 163) – is no longer sufficient for the embryo which breaks off the membranes tying it to the uterus and is subsequently birthed. Similarly, if there is insufficient menstrual blood, and so nourishment, the child would be born before the expected delivery date. However, according to C's report, not all of the stored menstrual blood is employed and so, during the first period of the foetus' formation, which lasts 30 days for the male and 42 for the female, an accumulation of the residual blood which must be expelled after the delivery (lochial purgation) is created. Other physicians at the time made the link between milk and menses: one other Hippocratic author argued that in order to make the flow come it is necessary to place a cupping-glass to the breasts ('Epidemics' 2. 6, 16; cf. 'Aphorisms' 5. 50). He also stated that 'the period of milk is the brother of the period of menses' ('Epidemics' 3, 17). According to another physician, 'if the woman who does not give birth nor has ever given birth has no milk, she lacks menses' ('Aphorisms' 5. 39).

Some conclusions on Hippocratic gynaecology

It is undeniable that Hippocratic medicine was largely indebted to the popular medicine of its time, but it is equally true that the authors of some of the treatises in the *corpus* used more explicitly scientific models. In the case of menstruation, there are two possible explanations for the differences between their explanations and the claims made by modern science: the limits of technology and the cultural restrictions imposed on the gathering of evidence. Greek physicians did not have instruments capable of analysing inner organs and were not allowed to dissect corpses. Greek physicians founded their theories about inner anatomy on what human bodies produced and what was visible, following the principle, fixed by the Pre-Socratic scientist Anaxagoras, that 'phenomena are an instrument to see what is invisible' (Diller 14–42). In this context, menstruation was an interesting and mysterious fact, which compelled the physicians to theorise a different physiology for female patients.

Thus they attributed much importance to the observation of the flow. The phenomenon presented a basic regularity and anomalies were probably first experienced as a problem by the patient and then interpreted as a symptom of disorder by the physician: in this context, the fear of amenorrhoea (suppression of the menses) can be justified. Nevertheless physician C and his colleagues would have used any available instrument to solve the situation ('Prorrheticon 2' 24). In fact, the aim of a remarkable number of prescriptions, variously attributable to A and B, was to encourage 'evacuation'. Hippocratic physicians were aware of all the officinal plants which have expulsive properties, including mercury (Dierbach 54–5; French 81 n. 16; Grot 94), artemisia (Berendes 234; Dierbach 181) which is still claimed to evacuate delayed or absent menstruation, ivy (Dierbach 83; Grot 119), and the most effective and dangerous, hellebore, the gynaecological use of which was widespread (Gazza 81; Grot 97–8; Scarborough 361–2; Krug 117–18).

When the patient was a woman, therefore, the physician immediately observed the menstrual cycle. This explains why physician B created a descriptive model of female diseases, in which the first specified datum concerns characteristics of the flow (amount, quality, frequency), sometimes accompanied by the recording of the male seed's behaviour after coitus (above all, whether or not it is received in the womb). Judging from the prominent position it holds in each chapter, this type of model was extremely important for the physician. The behaviour of menses and of male sperm could be observed and therefore allowed a possible diagnosis: 'menses sometimes disappear, in some women first they flow and then stop and they are not natural, but always ill-smelling and less abundant than before and there is no seed in this period' ('Feminine Diseases' 141. 314, 10–12). Menstruation is also the basis of physician B's model for treatment of the patient who cannot become pregnant. The physician must determine the preponderant humour in menses and therefore the type of purgation required. If the flow, which has collected on rag, spread on an ash layer and then dried, is full of mucus then phlegm is preventing conception. If the flow is reddish and rather black then bile and saltiness are preventing conception ('Feminine Diseases' ch. 11; cf. Thivel 198). Still on the basis of the flow, physician B provided indications about the days on which a woman would be fertile. In one chapter he advised intercourse during menstruation, but towards the end of the flux ('Feminine Diseases' 11. 46, 5); elsewhere he states that: 'she must have intercourse with a man ... at the end of the menstrual flow or at the beginning: but it is better when it is at the end' ('Feminine Diseases' ch. 17; cf. 'On Barren Women' 221; 'On Generation' 53; 'On the

Eight-Month Foetus' 13). Physician C's statements are similar but more theoretically sophisticated: 'at that time the mouth of the uterus is widest and it is rigid after menses and the vessels attract the sperm, while in the preceding time the mouth of the uterus is more closed and the veins, being full of blood, do not attract the sperm to the same extent' ('Feminine Diseases' 24. 64, 1–5; 'On Barren Women' 213. 412, 13; cf. 'On Generation' 15). In modern medicine, the days on which menstruation begins or end are the least fertile.[12] In conclusion, it must be inferred that the physicians of the *corpus Hippocraticum* regarded the condition of menses as crucial to general health. As time passed, their claims about regular and irregular flows were adjusted in the light of new evidence. As these physicians had no notion of hormonal processes as the basis of menstruation, the physiological model carried the weight of explaining the event, an otherwise inexplicable phenomenon. That said, the high standard of scientific explanation and practice reached by the Hippocratic physicians must be acknowledged and common judgements about their bias must be revised.

Notes

1. I use the term 'menses' as the most literal translation of the two terms – *epimēnia* and *katamēnia* – used in the *corpus*, both of which are plural.
2. All translations are my own.
3. I attribute chapter 19 of 'Feminine Diseases' to stratum B for its insistence on the counting of the days, which are defined in connection with the menstrual cycle (see 10. 40, 14), as well as for its relationship with chapter 179 (19. 58, 11 is the same as 179. 362, 13).
4. The attribution of chapter 37 to A is based on several observations: the patient is immediately purified from humour excesses; rods called *molubdoi* are used (Preus 248–9); and, indications of the therapy's duration are provided (Grensemann, 2: 42–3).
5. See also Dean-Jones, 'Bleeding' 179–81; *Bodies* 88–9.
6. Chapter 57 of 'Feminine Diseases' must be attributed to A as a typical expression of his is used in order to indicate the areas of the body which are subject to stress by the disease (57. 114, 13–14). Moreover, the chapter suggests the milkcure and, as an emetic, the decoction of lentils, mixed with hellebore ('Feminine Diseases' 143. 316, 7; 146. 322, 8–9).
7. The chapter can be attributed to author B because it proposes the same regime as 110. 236, 4–238, 3; 238, 7–10 and 112. 240, 13–242, 8.
8. In chapter 1 of 'Feminine Diseases' it is stated that there is a difference between a woman's and a man's flesh. The latter, being rougher, is less spongy and soft, so it draws fat from the digestive apparatus more slowly (Hanson, 'Assumptions' *passim*; Manuli 188; Hanson, 'Conception' 37–8; Dean-Jones, *Bodies* 55–7).
9. For example, see King on physician C ('Bleed' 112–13).

10. On the connection of menstruation with urine and faeces, see Diepgen 127–8.
11. Amenorrhoea is part of the symptomatology of some diseases which are not gynaecological as such (cf. 'Aphorisms' 6. 29; 'Prorrheticon 2' 7). Menstruation is judged as a positive event for a sick woman (cf. 'Epidemics' 5. 1, 12; 'Coan Prognoses' 148; 'On Illnesses 1' 7).
12. Fertility is highest approximately in the middle of the menstrual cycle although pregnancy is possible during menstruation if a woman has an irregular cycle of 14 days, in which ovulation takes place during the flow itself.

Works cited

Andò, Valeria. *Ippocrate. La Natura Della Donna*. Milano: BUR, 2000.

Aristotle. *Opera Omnia*. Paris: A. Firmin-Didot, 1878–84.

Berendes, Julius. *Die Pharmacie bei den alten Culturvölkern*. Halle: Tausch & Grosse, 1891.

Dean-Jones, Leslie. 'Menstrual Bleeding According to the Hippocratics and Aristotle'. *Transaction and Proceedings of the American Philological Association* 119 (1989): 177–92.

——. *Women's Bodies in Classical Greek Science*. Oxford: Oxford University Press, 1994.

Di Benedetto, Vincenzo. *Il Medico e la Malattia. La Acienza di Ippocrate*. Torino: Giulio Einaudi, 1986.

Diepgen, Paul. 'Die Frauenheilkunde der alten Welt *Handbuch der Gynäkologie*'. Ed. Walter Stoeckel. München: J. F. Bergmann, 1937.

Dierbach, Johann Heinrich. *Die Arzneimittel des Hippokrates, oder Versuch einer Systematischen Aufzählung der in allen Hippokratischen Schriften Vorkommenden Medikamenten*. Heidelberg: K. Groos, 1824.

Diller, Hans. '*Opsis adelon ta phainomena*. Der griechische Naturbegriff'. *Hermes* 67 (1932): 14–42.

French, Valerie. 'Midwifs and Maternity Care in the Roman World.' *Helios* 13. 2 (1986): 69–84.

Gazza, Vittorino. 'Prescrizioni mediche nei papiri dell'Egitto greco-romano. 2'. *Aegyptus* 36 (1956): 73–114.

Goltz, Dietlinde. *Studien zur Altorientalischen und Griechischen Heilkunde. Therapie-Arzneibereitun-Rezeptstruktur*. Wiesbaden: Steiner, 1972.

Grensemann, Hermann. *Knidische Medizin*. 2 vols. Berlin: Gruyter, 1975–87.

Grot, Rudolf von. 'Über die in der hippokratischen Schriftensammlung enthaltenen pharmakologischen Kenntnisse'. *Historische Studien zur Pharmakologie der Griechen, Römer und Araber*. Dorpat: Schnakenburg, 1887.

Hanson, Ann Ellis. 'Anatomical Assumptions in Hippocrates' *Diseases of Women* I, 1: My Mother the Earth'. *Society for Ancient Medicine Newsletter* 8 (1981): 4–5.

——. 'Conception, Gestation and the Origin of Female Nature in the Corpus Hippocraticum'. *Helios* 19 (1992): 31–71.

——. 'Hippocrates: *Diseases of Women* 1, Translated and with a Headnote'. *Signs* 1.2 (1975): 567–84.

——. 'Phaenarete: Mother and Maia.' *Hippokratische Medizin und antiken Philosophie-Verhandlungen des VIII. Internationalen Hippokrates-Kolloquiums*. Ed. Renate Wittern and Pierre Pellegrin. Hildesheim: Olms, 1996. 159–81.

——. *Studies in the Textual Tradition and the Transmission of the Gynecological Treatises of the Hippocratic Corpus*. Philadelphia: Pennsylvania University Press, 1971.

Hippocrates. 'Aphorisms'. Vol. 4 of *Oeuvres Complètes*. Trans. Émile Littré. 10 Vols. Paris: Baillière, 1844. 458–608.

——. 'Coan Prognoses'. Vol. 5 of *Oeuvres Complètes*. Trans. Émile Littré. 10 Vols. Paris: Baillière, 1846. 5: 588–732.

——. 'Feminine Diseases'. Vol. 8 of *Oeuvres Complètes*. Trans. Émile Littré. 10 Vols. Paris: Baillière, 1853. 10–406.

——. 'On Air, Water and Plants'. Vol. 2 of *Oeuvres Complètes*. Trans. Émile Littré. 10 Vols. Paris: Baillière, 1840. 12–92.

——. 'On Barren Women'. Vol. 8 of *Oeuvres Complètes*. Trans. Émile Littré. 10 Vols. Paris: Baillière, 1853. 408–62.

——. 'Epidemics 2'. Vol. 5 of *Oeuvres Complètes*. Trans. Émile Littré. 10 Vols. Paris: Baillière, 1846. 72–138.

——. 'Epidemics 5'. Vol. 5 of *Oeuvres Complètes*. Trans. Émile Littré. 10 Vols. Paris: Baillière, 1846. 204–58.

——. 'On Generation, On the Nature of the Child, On Illnesses 4'. Vol. 7 of *Oeuvres Complètes*. Trans. Émile Littré. 10 Vols. Paris: Baillière, 1851. 470–614.

——. 'On Illnesses 1'. Vol. 6 of *Oeuvres Complètes*. Trans. Émile Littré. 10 Vols. Paris: Baillière, 1849. 140–204.

——. 'On Illnesses 2'. Vol. 7 of *Oeuvres Complètes*. Trans. Émile Littré. 10 Vols. Paris: Baillière, 1851. 8–114.

——. 'On Regimen'. Vol. 6 of *Oeuvres Complètes*. Trans. Émile Littré. 10 Vols. Paris: Baillière, 1849. 466–662.

——. 'On the Eight-Month Foetus'. Vol. 7 of *Oeuvres Complètes*. Trans. Émile Littré. 10 Vols. Paris: Baillière, 1851. 452–60.

——. 'On the Nature of the Woman'. Vol. 7 of *Oeuvres Complètes*. Trans. Émile Littré. 10 Vols. Paris: Baillière, 1851. 312–430.

——. 'On the Seven-Month Foetus'. Vol. 7. of *Oeuvres Complètes*. Trans. Émile Littré. 10 Vols. Paris: Baillière, 1851. 436–52.

——. 'Prorrheticon 2'. Vol. 9 of *Oeuvres Complètes*. Trans. Émile Littré. 10 Vols. Paris: Baillière, 1861. 6–74.

King, Helen. 'Bound to Bleed: Artemis and Greek Women'. *Images of Women in Antiquity*. Ed. Averil Cameron and Amélie Kuhrt. London: Routledge, 1993. 109–27.

——. 'Sacrificial Blood: the Role of the Amnion in Ancient Gynecology'. *Helios* 13. 2 (1986): 117–26.

Krug, Antje. *Heilkunst und Heilkult*. München: Beck, 1985.

Lonie, Iain M. *The Hippocratic Treatises*. Berlin: De Gruyter, 1981.

Manuli, Paola. 'Donne mascoline, femmine sterili, vergini perpetue. La ginecologia greca tra Ippocrate e Sorano'. *Madre Materia. Sociologia e Biologia della Donna Greca*. Ed. Silvia Campese, Paola Manuli and Giulia Sissa. Torino: Boringhieri, 1983. 147–204.

McLaren, Angus. *A History of Contraception: from Antiquity to the Present*. Oxford: Blackwell, 1990.

Preus, Anthony. 'Biomedical Techniques for Influencing Human Reproduction in the Fourth Century B.C.' *Arethusa* 8 (1975): 237–63.

Scarborough, John. 'Theophrastus on Herbals and Herbal Remedies'. *Journal of History of Biology* 11 (1978): 353–85.

Thivel, Antoine. *Cnide et Cos? Essai sur les doctrines médicales dans la Collection hippocratique.* Paris: Belles Lettres, 1981.

Wenskus, Otta. 'Vergleich und Beweis im hippokratischen Corpus'. *Formes de pensée dans la Collection hippocratique. Actes du IV Colloque International Hippocratique.* Ed. François Lasserre and Philippe Mudry. Genève: Droz, 1983. 393–406.

2
Menstruation in Aristotle's Concept of the Person

Gabrielle Hiltmann

Western philosophy has always been concerned with metaphysical systems and metaphysical ideas. In this chapter I highlight that there are, and have been, diverse approaches to metaphysics and, within Western philosophy, many different metaphysical systems. Plato, for instance, developed his philosophical system against the background of Greek myth and this literary approach to metaphysics differed from that of Aristotle, who attempted to integrate what we would today call physics and biology into a general philosophical worldview. Metaphysics, as the general study of being, intersects with the social sciences, arts and even jurisprudence, but the way in which each is conceived as intersecting depends on the situation of the philosopher. As the location of the philosopher influences the way in which s/he approaches metaphysics, her/his understanding of aesthetic, social, ethical, legal and political matters will vary. Because the situation of a philosopher is political and social it can be argued that metaphysical and non-metaphysical realms interact and even mutually constitute one another depending on the particular nature of the interaction. I term the pattern developed by a philosopher to interlink the metaphysical and non-metaphysical a scheme. Schemes have various objectives. Firstly, they give a specific – scientific or technical – form to a relation. Secondly, they give a specific orientation in a complex field. Finally, they may serve as a model that becomes normative. Philosophical schemes conceptualising the relationship between men and women tend to activate two, intermingled, paradigms: nature and culture, body and mind/soul. These particular schemes combine all three objectives and, specifically, relate scientific biological claims to social (normative) functions.

Aristotle's thought provides a remarkable example of this mutual interaction of multiple constituents: physics, what we call today biology (especially the reproduction of animals), ethics, logic, and metaphysics. The aim of this chapter is to show the complex conjoining of these different realms in Aristotle's concept of the person. His vision of the person took up his metaphysical assertion that every being is formed matter. To understand Aristotle's model of how the person, as a being whose body is informed by human shape, is generated, it is necessary to examine in detail the role of menstruation which, according to his perspective, was one of the main factors of human reproduction. From certain metaphysical speculations about women, and their lack of 'form-giving force', Aristotle developed a double conception of the relation between women and men. He linked a monist one-sex scheme, in which men and women are basically the same, with a dualist two-gender scheme, where men and women should occupy distinct and hierarchically-organised social roles. Interestingly, both conceptions belong to an over-all 'logic of the same', which shapes Aristotle's concept of the person (Agacinski 126–9; Shildrick 28). The goals of this chapter are, firstly, to understand how women and men can be the same yet in a hierarchical relationship both in reproduction and within the polity and, secondly, to understand the ideological implications of Aristotle's concept of the person. Aristotle produced a complex conceptual apparatus designed to integrate men and women into a monist conception of a person but, despite this, 'woman' remains outside the Aristotelian scheme. In this exterior position she plays a fundamental and unacknowledged role, as a sort of hinge around which the logic of the same can function smoothly.

In the context of his general philosophy of substance, Aristotle conceived all beings as a specific relation of form and matter – a metaphysical concept of the person (Föllinger 126 and 139 ff.). He saw form as a general force that categorises matter (*Metaphysics B* 999 b 15–17, and *Metaphysics Λ* 1069 b 33–1070 a 2). *Hyle*, the word for matter, means 'brushwood', 'material', 'sediment' and 'excretion'. *Eidos*, the word for form, means the visible shape (*De Anima* 412 a 9–10). He saw the inter-growth of form and matter as essential for procreation, and saw one parent contributing matter – the body of the child – and the other filling this matter with form by giving it the soul (*De Anima* II 412 19–20). He found it obvious that it is woman who contributes matter, and man who shapes this matter with human form. The matter of sperm had no impact on procreation; conception could even happen without sperm. Aristotle understood that the material part of the energy of male sperm dissolves and evaporates (*Generation* 737 a 5–15). Generation implies a dichotomous starting point. This seemingly simple connection and distribution of two contributions to procreation was integrated in a complex metaphysical theory. Aristotle's vision of the person interlinked physiological and immaterial constituents. He understood that from conception onwards, form and matter, soul and body, undergo a process of 'intergrowth' that attains different levels. The nutritive and procreative soul possessed by plants is fundamental for all the other souls. The perceptive and sensitive soul develops on a higher level, and this desiring and mobile soul causes movement within animals. Last, the highest, rational, soul is possessed only by man (*De Anima* II 414 a 34-II 414 b 18). Because all lower souls are included in the higher souls, man has all of them. Despite their mutual relation, each type of soul 'behaves' differently. The physiological soul is incorporated, meaning that it dies along with the body, whereas the rational soul can be separated from the body, and is divine and immortal (*De Anima* 412 b 4–9, *De Anima* II 413 b 24–7, *Generation* II 736b 20–5). The rational soul is meta-physical (*De Anima* 413 a 6-b 24 ff.; 430 a 22; *Metaphysics* Γ 1070 a 26). This graded system of souls was used by Aristotle as a scheme to distinguish the roles and positions of women and men in reproduction, as well as in their capacity to contribute to public affairs. Aristotle's ideas concerning women's and men's roles within sexual reproduction can be further clarified by examining his claims regarding the function of menstrual blood.

Menstrual blood and sexual differentiation

For Aristotle, life meant participation in female and male qualities. For many animals, this participation was distributed between two types of *the same species* (*Generation* II 732 a 10). This distinction, however, presented Aristotle with a conceptual problem: as men and women interact in reproduction they must both belong to the same species, as beings belonging to different species cannot – or only exceptionally as in the case of mules, where horses and donkeys interact in reproduction – produce common offspring. But how is it conceivable that there are two different types of beings in one and the same species? Aristotle constructed a solution to this problem based on a complex physiological and physical theory. For him, the heart was the place of sexual differentiation. The heart ferments blood, which meant, for Aristotle, that it adds motile energy. Drawing on his argument that due to their rational capacities men's souls are more active, he concluded that they add more active energy to their blood, transforming it into sperm. As sperm can leave the body with great energy, he further concluded that sperm has more force than menstrual blood. It was equally evident to Aristotle that women's souls have less energy, which means that, because blood cannot be sufficiently fermented, it becomes menstrual blood, which leaves the body with little energy. Aristotle concluded that menstrual blood has no vital force. Nevertheless, he stated that sperm and menstrual blood belong to the same category of bodily fluids: both are excretions and have reproductive functions, so differ only in warmth, energy and quality.[1] From these metaphysically-informed biological observations Aristotle deduced that it is only male sperm which contains all of the souls, including the highest, rational, soul, which shapes the child. In contrast, the lower fermentation of menstrual blood was an index that it contained only the nutritive and reproductive soul (Lange 1–15). Aristotle contrasted men and women through women's supposed inability to engender the non-physical first and forming movement (*Generation* IV 765 b 5–10). As menstrual blood figured as a sort of sperm of lesser quality, participating in generation, albeit with less warmth and less energy, but nevertheless contributing the nutritive and reproductive soul to the new being, Aristotle considered it to be male.

Aristotle thus explained role-differentiation in reproduction as a difference in force, which he grounded in the metaphysical power of form. He believed that this also explained the differing capacities of women and men to transmit the shape of the species in generation. By using a gradated scale of temperature, Aristotle distinguished two beings with different energy levels inside the same species. However, if the difference between women and men was not a question of form – as both belong to the same species – it might be a question of matter: sex differences might be constituted by sexually-differentiated matter. Again, Aristotle's answer to this problem was a monist one. As the normal shape of a child is male, matter produced by women must be capable of producing the male sex. In order to be able to do this, menstrual blood must be intrinsically male. This means that there is only one matter and this matter is male. Interestingly, Aristotle did not consider the possibility of neutral matter being able to develop into female and male shapes. His concept of reproduction, and of the relation between sexes, is a striking example of an idiologic of the same. If, for Aristotle, the normal child was male, why were girls ever produced? He explained that it is only by accident that girls are formed or rather de-formed, since they have no penis and less heat. The lack of male sex is a visible sign of woman's monstrosity. This accident happens if the procreating man is too weak as the result of being too old or too young, or if the conception is influenced by cold winds. In these cases man cannot impose his form on the menstrual blood. So women are an accident of nature, but, as mammals reproduce themselves through two different beings contributing the form and matter of the offspring, are also a useful and even necessary one.

Aristotle's remarks on the procreation of the human species were part of his more general investigation into the reproduction of animals, and especially of mammals, in the book *Generation*. Several experts on Aristotle's philosophy considered even in the twentieth century that his conception of reproduction was – despite some minor errors – generally correct (Randall 78). A. L. Peck says in a 1949 foreword to his translation of Aristotle's *Generation*, 'this [Aristotle's comparison with fermentation of cheese] is a remarkable intuition of the essential rôle played by fermentation in embryonic development.' A.L. Peck is the translator of the classical edition of Aristotle's works, and systematically places the word 'male' before the word 'female' even though in the original 'female' often comes before 'male'. Nevertheless there are also critical voices speaking of Aristotle's epistemological blindness regarding the inferior status of women in his biological writings (Byl 210–22).

Aristotle's 'logic of the same' regarded the male shape, which integrates the rational soul, as the standard shape of the human species. The implication is that women as human beings are conceptually not just the *same* as men but *actually* men, distinguished from men only by their lack of form. Their deviance from this standard human form is considered to be monstrous. Aristotle even posited that the first departure towards the generation of a monster is the generation of female instead of male offspring (*Generation* 767 b 5). As the founder of classical formal logic, it is no surprise that it is possible to find a classical conclusion in Aristotle's texts concerning the monstrous status of woman. By bringing together different sentences that can be found in the *Generation of Animals*, it is possible to construct the following syllogism: Woman is by nature a deformed, castrated and impotent man (*Generation* 775 a), a deformed being is a monster (*Generation* 769 b 30), therefore a woman is a monster (*Generation* 767 b 5). What assumptions are implied by this argument? In Aristotle's view, every being that deviated from its normal shape was a monster. As the normal shape of the human was male, woman, being a deviant form of man, was a natural monster with a status even beyond that of eunuchs' 'artificial' monstrosity.

How can the status of these different deviant forms from the standard male human form be understood? Based on Aristotle's text it is possible to construct a one-sex scheme with different grades: The man who is full of form-giving force is in the highest position. Men who are weak or insane are below them. In the third position are the intersexed, both male and female. The fourth position includes eunuchs, without male sex by castration. The lowest position is that of woman, who – because of her contribution of matter in reproduction – is without male sex by nature. It is crucial to note that in the one-sex scheme, the morphological difference between woman and man, linked to the absence or presence of the penis (which is a contradictory opposition), is situated in an overall scheme where women and men are in a gradated relation. Female and male living beings belonging to the same species are not completely different (Deslauriers 142 f.). They differ only in view of the sexual organs and their non-equal contribution to procreation. The female body participates in the form of the human species (Aristotle *Metaphysics* I 1058 a 29–31 and *Generation* I 716 a 20–30). In the gradated one-sex scheme of the human, the difference between man and woman is not an essential difference.[2] Menstruation therefore had three functions in Aristotle's concept of the human. First it was the necessary counterpart to the form of the species transferred in generation. Second – as menstrual blood is male and thus contributes male matter to reproduction – it confirms the monist concept of the person. And third, as it ranks lowest of the gradated products of the heart's fermentation, it explains the gradated difference between woman and man with reference to the principle of a single form shaping the species.

Ideological implications of Aristotle's concept of the person

A close reading of Aristotle's book, the *Generation of Animals*, reveals traces of generalisations about masculinity. The extensive, yet speculative, discussion of the nature of male sperm and its different physical states, in view of the distinctions of warm and cold, fluid and solid, fluid and frozen, can be read as an indication of the hedonistic and (auto-)erotic importance of ejaculation for man (in this case, for Aristotle). Aristotle's concept of the human as male was extrapolated from his own physiological experience, thus preventing a recognition of a qualitatively different function for menstrual blood in the process of reproduction.[3] Menstrual blood was conscribed within a notion of homology to male sperm where it ranks as an excretion of minor quality. The Greek word *sperma* does not only mean 'sperm' but also 'race', 'origin' and 'descent', revealing a conviction that procreation of the human species is almost exclusively a matter of male procreative force for the reproduction of men.[4] The idea that the life-engendering principle of form lies completely in the male sperm is an example of a theory that takes the male body as its starting point. This theory intersected with Aristotle's metaphysics, which it confirmed, and by which it was in turn ideologically supported. It is noteworthy that in Aristotle's philosophy, the thesis of female inferiority was not physiological but ideological. Woman is not inferior because of her physiological difference. Aristotle characterised the essence of female inferiority by the supposed incapacity to engender the non-physical first and forming movement (*Generation* IV 765 b 5–10). The physiological difference is a consequence of a metaphysical principle. Her soul – lacking the principle of reason – is weaker than that of man. This is the reason why, Aristotle understood, she cannot impose herself against man – either in reproduction, or in public life.

Whereas Aristotle's conception of women and men in reproduction arranged them hierarchically within a monist scheme, in his political thought, on the other hand, the social relationship of women and men was *dichotomous*. Their different social roles are conceived as an analogy of the aristocratic form of government. Both partners reign in their clearly separated domains, and their capacity to act is different: woman is active in the house, keeps the household, and is responsible for reproduction and the education of infants; man is active in the public domain, especially in the agora where affairs of the city are dealt with. The relation between women and men in the home is characterised as friendship which, in Aristotle's thought, could be founded on three different grounds. The first ground is interest. In the case of a married couple this concerns mainly the economic interest of good housekeeping. The second ground is sexual desire. Sexual desire in the relation between women and men ensures the reproduction of free citizens. And the third ground is mutual love for the other person. It is noteworthy that in all three forms of friendship the relation is not that of equal partners. Based on his metaphysically legitimated capacity of form, man is considered to be superior to woman. Friendship founded in superiority is based on an asymmetric equality. Because of the different dignity of the two partners, friendship must be a proportional relation, with each partner receiving what is owed to him or to her. This implies that man, who is considered to be the superior partner, has more rights, including the right to be loved more. Despite the possibility of an asymmetric equality of woman and man in the sphere of the house, woman is completely excluded from any participation in the public realm. The scene of the agora is dominated by a logic of the same based on homoerotic relations between older and younger citizens.

Homosexual relations between men and the exclusion of women from democracy

In ancient Greece homosexual relations between men were not only considered to be perfectly natural, they were, as Jean-François Lyotard states, also a constitutive element of Greek democracy. He describes the libidinal economy at the roots of the Greek democracy as a zero-sum game (155). The bisexuality of free citizens assured the reproduction of children and warriors necessary for the survival of the city. Free from their duties of reproduction, married citizens lived in homoerotic mentor relationships in the agora. These relations aimed at the education and the social, as well as the philosophical, integration of the younger men. In this constellation of sexual initiation into political life, the decency of the young men and the mentors' mindfulness for the sexual honour of the younger men were of primary importance for their further political careers. Decency was not linked to sexual abstinence. It was important that the young man reacted with dignity towards the approaches of older men and that they chose their lovers circumspectly (Foucault, *Use* 204 ff., 217 ff.). In Plato's *Symposion*, all the participants agree that, in the absence of legal prescriptions, adult men have to wait for a certain (spiritual) maturity of the boy before they engage him in sexual intercourse (181C–E).[5] The erotic encounter between men and boys also included platonic aspects, as becomes evident in Diotima's plea for homosexual relations with boys in Plato's *Symposion*. The quest for the idea of the good and beautiful, seen as the highest form of love, was realised in the (not exclusively) platonic dialogue between men, and especially in the educational dialogue with younger men. The adult lover was supposed to fertilise the beauty, goodness, and bravery of the boy by education, and the results of this educational intercourse were supposed to be of greater value than the generation of natural offspring, as the products of the platonic love between men – literary works, works of art or laws – were considered to be of greater importance, and less mortal than human children (Plato 209B–E). It is in this double sense of homoerotic relations that, in ancient Greece, relations between men were considered to be the most valuable relations. As a consequence of the homoerotic foundation of Greek democracy, free Greek women were excluded from participation in the democratic government and the development of democratic institutions. Their exclusion was metaphysically legitimated by their lack of form-giving force. Woman's exclusion was, on the other hand, the necessary condition for the development of a space of homoerotic interaction between men in the public realm.[6]

In legitimating the exclusion of woman from the public realm, Aristotle linked her exclusion from the political realm with his ideal of male superiority in procreation. Nevertheless, social discrimination against woman was not legitimated simply by understandings of physiological difference. To designate the male force in procreation Aristotle used the verb *kratein*, meaning 'to reign'. He also used the verb *kratein* to describe man's social authority in the space of the house, and justified this social distribution of power with reference to the greater virtue of man. Man can reign in the house as an absolute ruler in much the same way as he reigns in reproduction, as the one who transmits the human shape of the child. Man's capacity to reign is linked to his form-giving power. Again, Aristotle included metaphysical principles in his arguments. The inferior social status of woman is deduced not from her physiological difference but from her minor contribution to procreation. The specific being of woman in the metaphysical sense, and her *logos*, are weaker than those of man. For this reason, woman holds an inferior position in the gradated scheme of the person. The Greek word *logos* means at one and the same time 'rational capacities' and 'the capacity to speak'. In *Metaphysics Γ* Aristotle compared humans who are not able to speak in a logically correct way with plants (1006b 7–15, and 1008a 30f). He insisted in this text that it is necessary to avoid multiple senses when we say 'something of something' and that, to do this, we must acknowledge several logical principles: the law of the excluded middle, the law of non-contradiction, and the law of identity. Thanks to these logical instruments, the essence of a (scientific) object can be defined precisely. Somebody who does not respect these logical principles – especially the law of non-contradiction – does not say anything because he is unable to speak precisely; he, or especially she, is like a plant. It is not possible to converse with such persons. This incapacity to speak logically concerns of course children, and particularly women, whose souls Aristotle compares explicitly with plants due to the lack of form-giving force in reproduction (*De Anima* II 413 a 6-b 24 ff., 414 a 34 and 430 a 22).

As woman cannot speak in a logical way, she can neither participate in scientific research nor speak in the Agora. Aristotle does not say that woman cannot speak at all, but that what she says makes no sense. It does not fit into logical categories. For this reason, she is excluded from the public realm where political affairs are discussed and decided. Because of the lack of the capacity to form, as well as the capacity for rational language, woman does not have the force to impose herself against man, either in procreation or in the running of the state. Of course, the lack of logical capacities in a woman is considered by Aristotle to be a lack in her specific being as an inferior man, not a result of lack of education. The clear distinction between private and public realms and the clear distribution of female and male roles in ancient Greek society show a clear two-gender distinction (Arendt 28 ff.; Cavarero 14, 16). This dichotomous distribution of social roles stands in a certain tension with the one-sex scheme developed by Aristotle in his reflections on reproduction and in his anthropology. Here, although woman is an inferior and less powerful man she nonetheless qualifies as a man. She makes an inferior, but nevertheless necessary, contribution to procreation. In the public domain, however, woman cannot participate at all. Instead of a gradated division of labour between woman and man, there existed a clear-cut distinction between two spheres. The exclusion of woman from the public realm was justified by beliefs in her inability to speak within the bounds of sense because she lacks the rational, highest, soul that enables man to shape, in a coherent and non-contradictory way, the world and the society in which he lives. This metaphysical argument is the basis for, on the one hand, the gradated distinction of woman and man in the one-sex scheme of reproduction, and, on the other, the exclusion of woman from the public realm in the clearly two-gender scheme.

Mechanisms of argumentation in Aristotle's concept of the person

Aristotle's arguments concerning the relation between woman and man were complex. First he developed metaphysical principles such as the principles of first cause and first matter. The principles of form, movement, and energy that man contributes to reproduction are neither biological nor physical. They are the principles of first form, first movement, and first energy. All these principles are metaphysical principles. The distinction and interconnection of form and matter, as well as the idea of souls with different qualities functioning on a gradated scale, are all metaphysical ideas. To explain natural phenomena he developed physical and physiological laws and principles anchored solely in his initial metaphysical frame of reference. But it is also evident that these physiological laws and principles support the metaphysical framework, so constituting a coherent and stable system. Aristotle found evidence for his one-sex scheme in his biological research on the reproduction of animals, but this research itself was guided by his metaphysical conceptualisation that every living being is an intergrowth of form and matter. Reproduction is explained by a complex interrelation of physical and metaphysical arguments. The notion that woman contributes passive matter and man contributes active form to reproduction was part of a metaphysical system, but this was also explained using physical vocabulary. Sperm, for example, is supposed to shape menstrual blood in the same way that acid added to milk shapes it into cheese by separating the fat from the liquid. Here, Aristotle used a practical experience as an analogy to explain the interaction of form and matter. The two levels are regularly interlinked by physical (for example, the concept of cause and effect), physiological and metaphysical vocabulary as well as knowledge of cultural techniques (fermentation of milk to cheese). Aristotle's theory of reproduction both confirmed, and was ideologically supported by, his metaphysics. Further on, he linked his own dichotomous logical forms of judgement with his metaphysics, and in this way developed a close link between his logic, orientated towards eternal conceptual entities, and his concept of substance (*Metaphysics Γ* 1007 a 10–26).

'Ideological' has to be understood in the Greek sense of the word: *idos logein* means to 'say one's own, the personal'. An ideological theory is one that is limited by one's own experience, especially by one's own bodily experience – which of course is shaped against the background of a social practice.[7] Owing to Aristotle's fundamental incapacity to distance himself from his own body, and to imagine the female body as something distinct and independent, he was also unable to consider that the female body contributes something decisive and positive in procreation. This continual refusal to recognise distinctness results in his insistence on a sex-based hierarchy within a monistic scheme. Thus one can conclude that his vision of the relationship of woman and man does not allow for any acknowledgement of an alterity in sameness. Aristotle's definition of the essence of the human as male in the process of reproduction conjoined his dichotomous logic with his metaphysics of substance and his natural research. In turn it was ideologically supported by this construction. The blind spot in this multifaceted construction is 'woman'. Woman is, by definition, matter which is not entirely formed, since she is not a complete man. Furthermore, woman produces 'not yet formed' matter for reproduction. Since woman is necessary for reproduction, she is at the same time inside, and at the limits of, the Aristotelian system, which is constructed by contradictory or contrary oppositions via which he tries to define matter according to clear and unambiguous notions. 'Woman' is, at once, the border of the system, and the pivot around which Aristotle's metaphysics turns, almost without friction.

Notes

1. Concerning the dichotomous differentiation of female passive matter and male, active and divine energy or form, the transfer of the *eidos-hyle* conception of procreation was not always coherent. In some passages Aristotle considered an active contribution by women which, nevertheless, was in most cases negative. In a decisive passage of *Generation* he refuted in depth the two-sperm conception as developed by Hippocrates and his students (I 721 a 30–731 b 14). Aristotle stressed – based on the *eidos-hyle* conception – that it was only man who contributes the energetic first principle of form and that woman only contributes matter.
2. See Deslauriers 140 ff. (on metaphysics), and 147 ff. (on physiology).
3. Lloyd is especially critical about Aristotle's attempts to confirm his patriarchal vision by 'facts' (Lloyd 214 ff.).
4. 'This almost complete erasure of active female participation (in generation) mirrors Aristotle's view that the active spirit was an exclusively male attribute' (Shildrick 32).
5. In Roman society of the first and second century BCE, sexual relations between adult men and boys were considered natural. But nevertheless – in contrast to Greece – a law, *lex Scantinia*, protected free Roman boys from sexual harassment. Their teachers were bound to sexual abstinence and had to watch over their protégé as a father would. Because of the protection of free Roman boys, sexual intercourse often took place between adult men and slave boys (Foucault, *Care* 190 ff.).
6. For an excellent reflection on the impact of this beginning of Western democracies on the actual state of affairs, see Collin 14.
7. Aristotle's monist one-sex scheme was of far-reaching influence in Western societies, owing particularly to his status as a recognised authority for the education of physicians from antiquity until the end of the seventeenth century. Claudius Galen's medical vision of the human is an early example of the powerful effect of Aristotle's philosophy in medicine. In addition to the interpretation of Hippocrates' texts, the analysis of Aristotle's work was at the centre of Galen's second century CE research. In spite of Galen's major departures from Aristotle, under Aristotle and Galen's influence, physiology, general pathology and clinical medicine were considered under philosophical categories. Galen's students, for example, had to learn Aristotelian logic and philosophy. Because Galen was one of the acknowledged medical authorities throughout the Middle Ages, the Aristotelian understanding of the human as male was accepted until the seventeenth century (Shildrick 28; Temkin 65 ff.; and other chapters in the 'Science and Medicine' section of this collection). It was only in the eighteenth and especially in the nineteenth century that in the context of the change to a biologically-founded vision of the person a two-sex scheme became socially dominant (Laqueur 149 ff.).

Works cited

Agacinski, Sylviane. 'Le tout premier écart'. *Les Fins de l'Homme*. Ed. Philippe Lacou-Labarthe and Jean-Luc Nancy. Paris: Galilée, 1981. 117–32.

Arendt, Hannah. *The Human Condition*. Chicago: Chicago University Press, 1958.

Aristotle. *Analytika Posteriora*. Ed. W.D. Ross. Oxford: Clarendon, 1978.

——. *De Anima*. Ed. David Ross. Oxford: Clarendon, 1961.

——. *Generation of Animals*. Trans. A.L. Peck. Cambridge: Loeb Classical Library, 1990.

——. *Metaphysics*. Trans. Hippocrates G. Apostle. Bloomington: Indiana University Press, 1966.

Byl, Simon. *Recherches sur les grands traités biologiques d'Aristote. Sources écrites et préjugés*. Bruxelles: Palais des Académies, 1980.

Cavarero, Adriana. *In Spite of Plato*. Cambridge: Polity, 1995.

Collin, Françoise. *Le différend des sexes. De Platon à la parité*. Nantes: Editions Pleins Feux, 1999.

Cook, Kathleen C. 'Sexual Inequality in Aristotle's Theories of Reproduction and Inheritance'. *Feminism and Ancient Philosophy*. Ed. Julie K. Ward. New York: Routledge, 1996. 51–67.

Deslauriers, Marguerite. 'Sex and Essence in Aristotle's *Metaphysics* and Biology'. *Feminist Interpretations of Aristotle*. Ed. Cynthia A. Freeland. Philadelphia: Pennsylvania State University Press, 1998.

Föllinger, Sabine. *Differenz und Gleichheit. Das Geschlechterverhältnis in der Sicht griechischer Philosophen des 4. bis 1. Jh. v. Chr.* Stuttgart: F. Steiner, 1996.

Foucault, Michel. *The Care of the Self*. Vol. 3 of *The History of Sexuality*. 3 vols. 1976–84. Trans. Robert Hurley. London: Penguin, 1990.

——. *The Use of Pleasure*. Vol. 2 of *The History of Sexuality*. 3 vols. 1976–84. Trans. Robert Hurley. London: Penguin, 1985.

Lange, Lynda. 'Woman is Not a Rational Animal: On Aristotle's Biology of Reproduction'. *Discovering Reality: Feminist Perspectives on Epistemology, Metaphysics, Methodology, and Philosophy of Science*. Ed. Sandra Harding and Merill B. Hintikka. Dordrecht: Kluwer, 1983. 1–15.

Laqueur, Thomas. *Making Sex: Body and Gender from the Greeks to Freud*. Cambridge: Harvard University Press, 1992.

Lloyd, Geoffrey. *Magic, Reason and Experience. Studies in the Origins and Development of Greek Science*. Cambridge: Cambridge University Press, 1979.

Lyotard, Jean-François. *Libidinal Economy*. Trans. Ian Hamilton Grant. Bloomington: Indiana University Press, 1993.

Peck, A.L. Foreword. *Generation of Animals*. Cambridge: Loeb Classical Library, 1990. iii–ix.

Plato. *Symposion*. Trans. Tom Griffith. Berkeley: California University Press, 1989.

Randall, J.H. *Aristotle*. New York: Columbia University Press, 1960.

Shildrick, Margrit. *Leaky Bodies and Boundaries*. London: Routledge, 1997.

Temkin, Owsei. *Galenism: Rise and Decline of a Medical Philosophy*. Ithaca: Cornell University Press, 1973.

3

The Art and Science of Menstrual Balancing in Early Medieval China

Sabine Wilms

In this chapter, I shall discuss the explanation, diagnosis, and treatment of menstrual disorders in what can arguably be called the earliest gynae-cological text in Chinese medical literature, the second, third and fourth scrolls of Sun Simiao's seventh century CE *Beiji qianjin yaofang* (*Essential Recipes for Every Emergency Worth a Thousand Coins of Gold*).[1] Between the seventh and tenth centuries (the early medieval period in China) gynaecology emerged as a respected field of specialisation in Chinese medicine. As an aspect of this development, recipes for 'adjusting the menses', specifically medicinal teas and pills which addressed menstrual conditions, assumed a central role in male physicians' treatment of their female patients. The direct purpose of the prescriptions in this category was of course the alleviation of a large variety of menstruation-related disorders. But, more importantly, menstruation also came to be seen as a window into the hidden processes inside the female body. Thus, male doctors relied on a highly differentiated diagnosis, taking into account the timing, consistency, colour and volume of menstrual blood, as well as a wide variety of accompanying physical and psychological symptoms. The variations in a woman's menstrual cycle and related symptoms were then interpreted to indicate underlying root pathologies such as 'blood depletion' or 'cold in the uterus', the two main causes of menstrual blockage.

Perhaps even more importantly, menstruation also offered important clues about a woman's general condition in a healthy body. In traditional Chinese gynaecology, health was defined primarily in terms of the abundance and flow of the body's vital substances. For this reason, physicians focused their attention on Blood and on *qi*, the life-giving matter, breath, and energy at the core of traditional Chinese medicine, manipulated in such treatments as acupuncture, herbal prescriptions or *qigong*. While the diagnosis and treatment of male bodies focused primarily on *qi*, female bodies, by the time of the Song dynasty, had come to be associated with Blood, which led to the common sentiment that 'in women, one first regulates Blood, and in men, one first regulates *qi*' (Chen, Scroll 6, qtd. Furth 84). In this context, the significance of Blood in Chinese medicine needs to be explained briefly. I distinguish in this chapter between 'blood' in the strict biomedical sense and the capitalised 'Blood', which refers to the more inclusive traditional Chinese conception. In medieval Chinese gynaecological theory, Blood was intimately linked to female reproductive processes because its recurrent monthly discharge as menses marked the absence of pregnancy and, simultaneously, the ability to conceive. In addition, it was said to nourish the foetus during pregnancy and to rise up into the breasts and be discharged as breast milk after childbirth. In the case of pathological blockage and lack of circulation, it was believed to congeal into abdominal masses or accumulate in the uterus where it caused chronic infertility. Since Chinese physicians defined physical health in both men and women in terms of a regular and unimpeded circulation of blood and *qi* throughout the body, it was only natural that a woman's timely and plentiful monthly flow was interpreted as a sign of abundant health and balance.

In the context of female healthcare in China, the topic of menstruation is particularly interesting, because the (usually male) physician was forced to depend on the woman's own assessment and definition of her condition, rather than on his own 'objective' measurements and observations. The aetiology and treatment of any particular disorder were therefore always decided in a process of shared negotiation between physician and patient. Because of the overlap between delayed menstruation, seen as a pathological condition to be treated with formulas that would 'disinhibit the blood-flow', and a potentially unwanted pregnancy, the category of 'blocked menstruation' in particular offered women a space within which to negotiate and control their own reproductive health. The therapeutic similarities between treating a blocked menstrual flow and aborting a foetus were well-known to both the physicians and their patients and the line between the two could easily be crossed without ever having to articulate the condition of pregnancy.[2]

Unfortunately, original sources that describe the practices of female and/or illiterate popular healthcare practitioners in medieval China from their perspective have not been preserved. Nevertheless, sources from the following centuries suggest that the vast majority of women, whether from elite or peasant households, relied on an informal and extremely diverse network of healthcare providers, ranging from elite male literati physicians to religious exorcists and shamans, Buddhist monks and nuns, herbalists, midwives, itinerant medicine peddlers, pharmacists, and, most importantly, the knowledge of other women in the extended family. The content and categorisation of treatments in the gynaecological literature from this period offers important clues as to which areas of women's health were considered appropriate for a male physician's intervention. The contrast between a mere six per cent of the text devoted to obstetrics and roughly twenty five percent dedicated to conditions of menstrual and other vaginal discharge, for example, can be explained in this context. On the one hand, the hands-on manipulations and close physical involvement in the process of childbirth that were required in obstetrical complications marked this as a sphere dominated by female healers and midwives. The adjustment of the menstrual flow by complex medicinal decoctions and pills, on the other hand, became the core of women's treatment by male gynaecologists.

The 'women's recipes' of Sun Simiao's *Essential Recipes for Every Emergency*

In order to trace the origins of Chinese gynaecology and, closely related to this, the formation of theories about the causes and treatment of menstrual disorders, Sun Simiao's *Essential Recipes for Every Emergency Worth a Thousand Coins of Gold* constitutes an ideal point of entry. Composed around 652 CE, it represents what was apparently the most sophisticated level of medical theory and practice in the early Tang period. An encyclopaedic work with over five-thousand entries in the form of essays and recipes, it covers over 200 medical topics, including gynaecology, paediatrics, internal medicine, skin disorders, emergency medicine, dietary therapy and macrobiotic hygiene, vessel therapy and pulse diagnostics, and acupuncture and moxibustion (the burning of plant-derived moxa as a counterirritant). It is representative of a type of medical literature called *fangshu* or 'recipe texts'. While *fang* is often rendered as 'prescriptions' or 'formulas' in later medical texts, it refers, in this early context, not only to medicinal prescriptions, but also to instructions for acupuncture points, physical manipulation, rituals, and household advice, and should therefore be translated more accurately as 'recipes'. Revealing the great value which the male author placed on women's health, the section on 'women's recipes' is found at the very beginning of the text in scrolls two to four, after a general introduction in scroll 1. It comprises about 10 per cent of the entire *Essential Recipes*.

In the introductory essay to the gynaecological section, Sun Simiao emphasises the centrality of women's health for 'gentlemen engaged in the art of nurturing life' (Sun, scroll 2, introductory essay). At the very beginning, he states that 'women's disorders are ten times more difficult to treat than men's'. This became one of the most often-cited sayings in Chinese gynaecological literature. The introduction continues: 'From the age of fourteen on [i.e. the standard age for the onset of menstruation], [a woman's] *yin qi* floats up and spills over, [causing] a hundred thoughts to pass through her heart. Internally, it damages the Five Organs; externally, it injures the disposition and complexion. The retention and discharge of menstrual fluids is alternately early or delayed, obstructed blood lodges and congeals, and the central passageways are interrupted and cut off' (Sun, Scroll 2, introductory essay). Even without understanding the intricacies of traditional Chinese medical explanations for the relationship between Blood, the internal organs, and the disposition, it is easy to see the central role played by the menstrual cycle in this author's view of the female body. With the exception of the introductory essay, however, the 'Women's Recipes' are predominantly a collection of treatments and do not contain much theoretical elaboration.

For another perspective on menstruation, we therefore need to turn to the actual recipes and their organisation, to see how menstrual conditions were diagnosed, interpreted, and treated in clinical practice. The section on 'women's recipes' is organised as such: scroll 2 covers conditions related to childbirth, namely fertility, pregnancy, obstetrics, and lactation; scroll 3 is devoted almost entirely to postpartum recuperation, followed by a short section on miscellaneous recipes, which treat such gendered conditions as dreams of intercourse with ghosts and structural abnormalities of the reproductive organs; and scroll 4 contains treatments for menstruation and vaginal discharge, preceded by a short section with prescriptions for boosting and supplementing women's health in general. At first sight, the size of the section on menstruation – strictly speaking only about eleven per cent of the 'women's recipes' – does not seem so significant, especially when compared with the large amount of space devoted to pregnancy and postpartum care. But a closer look at the actual content of the text complicates the picture considerably, as well as revealing the close connection between the aetiological categories of menstrual problems and vaginal discharge. Together, these comprise a full quarter of the 'women's recipes' and provided a cornerstone for the definition of women's bodies as gendered and categorically different from men's or, in other words, for sex dimorphism.

Menstrual disorders: major pathologies, symptoms, and treatments

Consisting mostly of complex prescriptions for medicinal decoctions and pills, the two sections on 'stopped menstrual flow' and 'irregular menstruation' certainly address the pathologies suggested by the chapter titles, but in no case are these seen as isolated symptoms. To clarify my argument, I quote in full the second prescription from the section on 'stopped menstrual flow':

> Dried Ginger Pills:
> For treating women [who suffer from] chills and fevers; emaciation and thinness; soreness and wasting disorder; inertia and sluggishness; propping fullness in the chest; heaviness and pain in the shoulders, back, and spine; hardness, fullness, and accumulations in the abdomen, potentially with unbearable pain stretching to the waist; pain in the lower abdomen; vexation and aching in the four limbs; reverse counter-flow in the hands and feet, cold reaching the elbows and knees, or with vexing fullness and depletion heat in the hands and feet so that she feels like tossing herself into water; extreme pain in the hundred joints; constant discomfort and suspension pain below the heart; alternating chills and fevers; nausea, profuse drooling, and salivating every time in response to salty, sour, sweet, or bitter substances; or an [appearance of] the body like chicken skin; stopped menstrual flow; discomfort and difficulty with urination and defecation; and eating without generating muscles:

> One ounce [c. 37.3 g] each of dried ginger, *Xiongqiong* lovage, huckatoe, niter, apricot pit, leech, horse fly, peach pit, black chafer, and wingless cockroach; two ounces each of Hare's Ear, white peony, ginseng, rhubarb, Sichuan pepper, and donquai.

> Pulverize the 16 ingredients above and mix them with honey into pills the size of parasol tree seeds. On an empty stomach, take three pills with fluid [three times a day]. If no effect is noticed, increase to a maximum of ten pills [per dose] (Sun, scroll 4, section 2, second recipe).

By analyzing the medicinal composition of this prescription, it is possible to read this recipe backwards. We can thereby gain valuable insights into the author's aetiological understanding of blocked menstruation in general, and of the symptoms listed for this recipe specifically. The standard usage as described in medieval Chinese *materia medica* literature classifies each of the ingredients for this recipe into the three categories of moving Blood, of warming the body, and of supplementing Blood.[3]

This recipe is intended to address the root aetiologies of a lowered body temperature, of Blood depletion, and of Blood congestion or lack of flow, as interconnected causes of stopped menstruation and its accompanying symptoms (the significance of this constellation of aetiologies will become apparent below). Moreover, the detailed and highly specific lists of indications at the beginning of the other recipes in the two chapters on menstrual problems show that these conditions were commonly associated with postpartum problems, with lumps, hardness and pain in the abdomen, and with vaginal discharge in a range of colours and severity:

Donquai Pills:
A recipe for treating women's concretions and knots below the navel, accompanied by stabbing pain as if being gnawed by worms or stabbed with an awl or knife; maybe with red or white vaginal discharge, with the twelve disorders [of abdominal masses], with aches and pains in the waist and back, and a menstrual flow coming now late, now early.

Herb Paris Pills:
A recipe for treating women's twelve disorders of postpartum concretions, vaginal discharge, and lack of offspring, all of which are caused by the cold *qi* of coolness and wind. Maybe the woman squatted over the privy without proper care and sat there a long time before she had waited out the hundred days of postpartum [seclusion] and had completely eliminated the malign blood left-over in the womb and network vessels, allowing dampness and cold to enter the inside of the womb; ... [for treating various pains all-over the body, especially in the genital area]; stopped menstrual period; or discharge that resembles rotten flesh or is clear, yellow, red, white, black, or other colors, or resembling bean juice; or inauspicious dreams and thoughts' (Sun, scroll 4, section 2).

These lists point toward a considerable overlap between the aetiological categories of menstruation and the following section on 'red and white vaginal discharge, collapse of the centre [life-threatening haemorrhaging], and [chronic vaginal] leaking'. Reinforcing this impression, a careful comparison of early editions of the *Essential Recipes* shows that different editions placed numerous recipes in either of these sections.[4] The confusion of this complex of interrelated symptoms and treatments might be surprising to a modern Western reader. But in early Chinese gynaecology, the aetiologies, indications, and therapies for menstrual problems and vaginal discharge are closely connected or even used interchangeably. By explaining this relationship in greater detail and placing it in the larger historical context, we can obtain a much clearer understanding of Sun Simiao's treatment of menstruation and his view of the female body in general.

Menstrual disorders, vaginal discharge, and the meaning of *daixia*

The most obvious link between the pathological categories of menstrual irregularities and vaginal discharge is found in the fact that they can both present as the identical symptom of vaginal bleeding, interpreted either as red vaginal discharge or as menstrual fluid. In addition to menstruation and vaginal discharge, lists of gynaecological conditions or symptoms found in early Chinese medical texts often include a third category of pathologies, namely abdominal masses. In the *Essential Recipes*, these are subsumed under the chapter on menstruation, but they are a common category in other contemporaneous literature. This triad of disorders forms the basis for the earliest references to women's disorders in Chinese medical literature. Moreover, it provides the key to the emergence of gynaecology as a professional medical specialisation in the centuries following the *Essential Recipes*. In the earliest extant reference to gynaecology from the third century BCE, the physician Bian Que is described as a *daixia yi*: a 'physician of (the area) below the girdle'. The context makes clear that this expression refers to a medical specialist in the treatment of women's conditions. In Chinese medical literature, the term *daixia*, 'below the girdle', has come to refer less commonly to female pathologies in general, but is more frequently employed as a technical term to refer specifically to 'vaginal discharge'. Thus, it illustrates the close association, in the early history of Chinese gynaecology, between women's bodies and pathological discharges from the vagina as the defining characteristic of female disorders. In the first medical text with a separate section on women's health, Zhang Ji's *Jingui yaolue* (*Essentials of the Golden Casket*) from the first or second century CE, women's disorders are explained as 'all being below the girdle' and 'due to depletion, accumulated cold, and bound *qi*, causing the various conditions of interrupted menstruation' (ch. 22). Elsewhere, the author stresses that women's disorders located in the upper and middle sections of the body are to be treated no different from men's. Therefore, it is only conditions 'below the girdle', specifically those related to the reproductive organs, menstruation, and childbirth, as well as abdominal problems, which required gender-specific treatment. The recipes in this text reflect a view of the female body as prone to depletion and invasion by wind or cold, pathogenic influences that would then impede the circulation of menstrual fluids. Thus, menstruation appears as a key element in this early attempt to define the female body, but in a way that is already much more evocative than a mere concern with infertility.

After a gap of several centuries, texts of the seventh and eighth centuries depict a far more sophisticated and elaborate conception of the female body in both theory and practice. Composed only decades before the *Essential Recipes*, Chao Yuanfang's *On the Origins and Symptoms of the Various Diseases* provides a detailed list of aetiologies specific to women in eight scrolls. In these, Chao laid the foundations for a literature of gendered treatments to support and boost the health of a female body that was endangered and weakened by reproductive processes and therefore in need of special protection and attention. He treats the symptoms of menstrual problems, vaginal discharge in various colours, abdominal masses, and infertility as distinct categories, but explains them with practically identical aetiologies: in all conditions, the ultimate cause is the physical taxation from childbirth. This damages *qi* and Blood, causes physical depletion, and leads to a subsequent invasion by wind and cold. These pathogens lodge in the uterus and can either cause the Blood to congeal there or injure the channels. In either case, they impede the circulation of blood and the descent of the menses. Moreover, when *qi* is depleted, and therefore unable to constrain and control blood, this pathology can turn into vaginal discharge: because of the damage to the channels, the blood mixes with filthy fluids in the uterus, forming vaginal discharge. In a direct application of the standard correspondences between the five colours and the five internal organs in traditional Chinese medical theory, the colour of the discharge gives important etiological clues since it indicates which internal organ is primarily affected. For example, red and white discharges are, in this scheme, related to heart and lung conditions respectively. However, they can alternately be interpreted as indicating the presence of excessive heat or cold. In this text, the reader is therefore faced with a plethora of disease categories, all of which are subsumed under the topic of *daixia*. They are unified by a common aetiology, which links malfunctions of the female body directly to taxation damage from their reproductive functions.

The organisation and clinical strategies of the 'women's recipes' in Sun Simiao's *Essential Recipes* were obviously influenced by the way in which Chao Yuanfang interpreted and categorised women's disorders. Like his predecessor, Sun identifies the variety of women's disorders with a root cause of taxation damage. This causes physical depletion, which allows pathogenic wind or cold to enter from the vagina and obstruct the healthy flow of blood. At first, this might merely result in delayed or obstructed menstruation. But in severe cases, the result is a total menstrual blockage, which in turn is the most common cause of infertility, abdominal masses, or vaginal discharge. While these symptoms are thus intimately connected in terms of their causation, a careful analysis of the actual prescriptions reveals a slight differentiation in their treatment. Vaginal discharge, on the one hand, is regarded as a chronic problem caused by general weakness. It is therefore treated with stabilising and tonifying drugs like limonite, dragon bone, ginseng, deer antler, donquai and white peony, in order to replenish the underlying deficiency and to strengthen the body in general. These prescriptions are often linked to reproductive functions, as can be seen from the warning in one recipe that 'widows and virgins may not recklessly take this,' and that, 'if she takes an excessive amount of this recipe, she will give birth to twins' (Sun, scroll 4, section 3, recipe for Dragon Bone Powder).

Menstrual conditions, on the other hand, are regarded as problems related to the circulation of fluids. As in the example of Dried Ginger Pills translated above, they are thus treated mostly with strong blood-moving drugs such as peach pit, rhubarb, leech, or horse fly, in conjunction with herbs for expelling pathogens like asarum or Hare's Ear. In later gynaecological literature, these are referred to as *tongjing yao*, literally 'drugs to bring on the menses/make the menses penetrate', a medicinal category that, coincidentally, is not explicitly distinguished from abortifacients. In addition to the positive association of menstruation with regularity and a healthy circulation of fluids in the female body, women's vaginal bleeding also plays an important role in the expulsion of the highly pathogenic substance called *e lu*, 'malign dew'. Constituting one of the most dangerous and tenacious pathologies for women, this refers to stale and rotting blood left over in the uterus after childbirth. It is associated with an endless list of chronic health problems for the rest of a woman's life if it is not eliminated completely and promptly after childbirth with numerous uterus-cleansing and blood-dispersing prescriptions.

Once again, the key symptom for the presence of 'malign dew' is blocked menstruation, to be treated with the above-mentioned combination of blood-moving and pathogen-expelling drugs. In contrast to the intended action of the vast majority of his own recipes in the sections on menstrual conditions, however, it has to be stressed that in his short essays Sun repeatedly warns against using harsh drugs. Instead, his theoretical discussions express a preference for stabilising and tonifying drugs to address the underlying aetiology of depletion on a long-term basis. Rather than weakening an already depleted body further by increasing the loss of fluids through strong expellants with immediately visible results, gentle strengthening prescriptions should eventually cure secondary symptoms of excessive discharge of blood and other fluids, referred to as '*daixia*' on the one hand and the lack of proper menstrual flow on the other.

The contrast between Sun's theoretical approach and the treatment methods reflected in the actual recipes shows that the *Essential Recipes* is situated at an important turning point in the development of gynaecology. Based on patterns of transmission in the history of recipe literature in China, it is likely that Sun Simiao's recipes were for the most part merely collected and synthesised by the author, while the essays most likely reflect his personal contribution to this text. Under the influence of medical authors like Chao Yuanfang and Sun Simiao, professional gynaecology progressed and moved away from drastic and heroic medical interventions with such immediate and visible results as bringing on the menses or stopping a haemorrhage, which ultimately only weakened the body further. What emerges in Chao's aetiologies, and Sun's essays, is a more refined awareness of the underlying causes of women's disorders and the need for long-term treatment aimed at restoring the female body's strength and balance. Based on this, male physicians began to recognise a need for women's gender-specific treatment, necessitated by the perceived weakness and vulnerability of the female body related to its reproductive functions.

Balancing and the bio-medical

In the centuries following the publication of the *Essential Recipes* and culminating in the composition of Chen Ziming's *All-Inclusive Good Prescriptions for Women* in 1237 CE, the gendered aetiologies and therapies of gynaecology can be summarised in the maxim: 'In women, Blood is the foundation', an expression cited in most gynaecological texts from the tenth century, but supposedly a quotation from an earlier, now lost, text on childbirth. As Charlotte Furth has stressed, during this period, 'medical authors appeared more concerned with female difference than either earlier or later in Chinese history' (60–1). A comparison of the major gynaecological texts from the seventh to thirteenth centuries shows a dramatic shift in the proportion of content related to menstruation and vaginal discharge.[5] Even a superficial analysis reveals clearly that menstruation came to replace vaginal discharge as the key symptom of gynaecological diagnosis in medieval China. As part of this development, the way in which physicians interpreted menstruation shifted from focusing on its weakening and destabilising effects to positively appreciating it as a natural and healthy function, essential for ensuring reproductive health through the periodic emptying and cleansing of the uterus. In this context, menstruation served as an evocative symbol for regular cosmic cycles, for a healthy circulation and flow, and for a balance of *qi* and Blood. Consequently, the retention of menstrual Blood and the absence of a regular flow became the central paradigm for female pathology, which then caused such secondary conditions as infertility, vaginal discharge, or abdominal masses. With growing confidence in their profession, physicians moreover began to diagnose and adjust women's menstrual cycles in order to actively prevent these secondary symptoms in their practice of *yue tiao*, 'menstrual balancing'.

This view of menstruation may be useful in engaging with Western biomedical interpretations. As Emily Martin pointed out in *The Woman in the Body*, the metaphors used in Western biomedical textbooks to describe the processes of menopause and menstruation are anything but objective and value-free. Rather, the association of menstruation with processes of hormonal 'decline', 'degeneration' of the *corpus luteum*, 'deterioration' or even 'denuding' of the uterine lining, or 'loss' of blood and 'lack' of fertilisation can be linked to cultural attitudes about production and the proper function of the female body. This loaded depiction of menstruation as a process of breakdown and deterioration can be usefully contrasted with descriptions of analogous but un-gendered processes. The regular shedding and replacement of the stomach lining, for example, is called a 'periodic renewal' and is related to the 'secretion' of mucus and the positive function of mucus as a 'barrier' to stomach acids (Martin 33). Early Chinese gynaecological texts, admittedly, lack such detailed and rich descriptions of women's physiological processes, but the care and specificity with which Sun Simiao's heirs discussed, observed, diagnosed, and adjusted women's menstrual cycles, depending on their timing, duration, amount, consistency, colour, and a host of associated physical and mental sensations, expresses an attitude toward the female body quite different from that of the contemporary West.

It is common knowledge that in medieval China's patriarchal and patrilineal culture a woman's role was defined predominantly by her ability to give birth to a large number of male offspring in order to perpetuate the husband's family line. But in the gynaecological category of 'menstrual balancing', Chinese medicine has always provided women with the opportunity to control their reproductive cycles without the need to address explicitly an unwanted pregnancy. As the *Essential Recipes* exemplifies, Chinese physicians' treatment of menstrual disorders was motivated not by the desire to objectify and control a female body that was regarded merely as a temporary vessel for the male seed. Quite differently, Sun Simiao's essays and recipes clearly speak of a humanitarian, if paternalistic, desire to protect a vulnerable female body that required careful medical attention because of the depletion and taxation of women's reproductive cycles.

Notes

1. All translations are my own.
2. See Bray 317–32.
3. The earliest and most reliable source for understanding standard medicinal usage in China is undoubtedly the *Shennong bencao jing* from the Late Han dynasty which was edited, annotated, and enlarged by Tao Hongjing in the early sixth century CE.
4. Most modern Chinese editions are based on the woodblock print edition in 30 scrolls which was revised and published during the Song period by a government-sponsored editorial team under the direction of Lin Yi. I have compared this to several manuscript editions that are likely to predate Lin Yi's revisions, most notably a manuscript titled *Xindiao Sun Zhenren Qianjin fang.*
5. Contrast, for example, *On the Origins and Symptoms of Various Diseases* (c. 605–16 CE) where menstruation constitutes two and a half per cent and vaginal discharge almost ten per cent of the total gynaecological section, with *All-Inclusive Good Prescriptions for Women* (1237 CE) where the category of vaginal discharge is incorporated as a subheading under menstrual conditions, constituting less than a quarter of this section or two per cent of the total. In another seventh-century text, vaginal discharge constitutes eight per cent of the total while menstruation is non-existent as a separate category, whereas, in another twelfth-century text, menstruation constitutes 10 per cent of the total, as opposed to one per cent for vaginal discharge.

Works cited

Bray, Francesca. *Technology and Gender. Fabrics of Power in Late Imperial China.* Cambridge: Cambridge University Press, 1997.

Chao, Yuanfang. *On the Origins and Symptoms of Disease by Master Chao [Chaoshi Zhubing Yuanhoulun].* Ed. Zhu Xinnian. Taipei: Guoli Zhongguo yiyao yanjiusuo, 1996.

Chen, Ziming. *Furen Daquan Liangfang [All-Inclusive Good Prescriptions for Women]*. Ed. Ying'ao Yu. Beijing: Renmin weisheng chubanshe, 1985.

Furth, Charlotte. *A Flourishing Yin: Gender in China's Medical History, 960–1665*. Berkeley: California University Press, 1999.

Martin, Emily. 'Medical Metaphors of Women's Bodies, Menstruation and Menopause'. *Writing on the Body: Female Embodiment and Feminist Theory*. Ed. Katie Conboy *et al*. New York: Columbia University Press, 1997. 15–41.

Sivin, Nathan. *Chinese Alchemy: Preliminary Studies*. Cambridge: Harvard University Press, 1968.

Sun, Simiao. *Beiji qianjin yaofang jiaoshi. [Prescriptions Worth a Thousand Coins of Gold for Every Emergency]*. Ed. Jingrong Li. Beijing: Renmin weisheng chubanshe, 1996.

Wilms, Sabine. *The Female Body in Medieval Chinese Medicine*. Taos: Paradigm, 2005.

Zhang, Ji. *Jingui yaolue [Essentials of the Golden Casket]*. Ed. Keguang Li. Shanghai: Shanghai kexue jishu chubanshe, 1991.

4

Flowers, Poisons and Men: Menstruation in Medieval Western Europe

Monica H. Green

When the seventeenth-century biblical scholars and linguists translating the Bible into English for James I in the early seventeenth century came to Leviticus 15, that litany of polluted states, they translated verses 19–33 regarding woman's 'issue of blood' quite literally. In most cases, they referred to menstruation (and perhaps also the lochial flow, the four- to six-week flow that follows childbirth) simply as 'her issue'. Yet twice they called it 'her flowers', a colloquial expression that had been in common use in England since at least the thirteenth century. Flowers denote youth, freshness, fecundity, beauty, and indeed that is how the King James translators used the term elsewhere in their masterpiece of English prose. But a woman, uniquely, could be 'sick of her flowers'. It is in this puzzling phrase that we see the conceptual legacy that medieval Europe bequeathed to the modern Western world. The positive image of 'flowers' lay at one end of a spectrum of attitudes towards menstruation. At its other extreme were images of menstruation as contaminating, foul, and even poisonous. Medical views of menstruation were largely neutral, though scientific ones tended to be more disparaging. Menstrual blood might have powers both healing and harmful, sometimes natural, sometimes magical. It was a sign of fertility and health, though also sometimes of disease. More remedies to provoke its flow would be found in collections of medical recipes than methods to suppress it. Christian authorities, reviewing the Levitican prohibitions, debated whether menstruation or the lochial flow should be used as a reason to keep women out of church. All these views about menstruation – medical, scientific, religious, and popular – might circulate in isolation from one another, or they might converge, producing such striking notions as the belief

51

that Jewish men 'menstruated'. To date, no evidence has been found that there were particular rituals to mark women's first menstruation. Nor did the cessation of menstruation, although recognised as a normal phenomenon of ageing, elicit much notice in its own right. Yet menstruation was never invisible in medieval Western Europe – it was always a symbol of female difference (even if it occurred in men), serving as a marker both of female fecundity and female physicality.

As the *Oxford English Dictionary* has it, the English usage of 'flowers' to designate menstruation derives from the French *fleurs*, which, in turn, is a corruption or confusion with the term *flueurs*, i.e. 'flowings' ('Flower').[1] The earliest French texts on gynaecology, which date from the mid-thirteenth century, used *fleurs* or *flors* as the normative term (Anon., *Secres de Femmes*; Hunt 2:111), although the usage can in fact be traced back to the Latin, with *flos* or *flores* used to refer to menstruation documented in the twelfth century. Previous to this, the Latin terms used were alternately *purgatio menstrualis* ('monthly purgation'), *fluor* ('flux', a rare term only found in the fourth/fifth-century writer Caelius Aurelianus), or, most commonly, *menstrua* (literally, 'the monthlies').[2] When Jerome was writing his definitive Latin Vulgate version of the Bible in the fourth century, he used the terms *sanguis menstrualis, fluxus sanguinis*, or *menstruus sanguis* (Lev. ch. 15). In fact, no 'flower' analogy is attested in any western gynaecological texts, from the ancient Hippocratic texts of the fifth and fourth centuries BCE, in which two of the words used to refer to the menses are *gynaikeia* ('the womanlies') or *katamenia* ('the monthlies'), to the eleventh-century writer Constantine the African, who used *menstrua* consistently when translating medical texts from Arabic into Latin. Rather than a corruption or confusion of the Latin term *fluor*, the term 'flowers' was probably a colloquialism used by women who were descended from the Germanic ethnic groups that spread across western Europe in the early Middle Ages.

In the Lombard area of southern Italy, an anonymous writer in the middle of the twelfth century composed a gynaecological text in two versions: a 'rough draft' called *Treatise on Women's Diseases*, and a more polished version produced slightly later, called *Conditions of Women*. The *Treatise on Women's Diseases*, which in general employs a more colloquial Latin, used the Latin *flos* ('flower') as the normative term for menstruation throughout the text. When the *Treatise* was revised into the *Conditions of Women*, the author changed all the instances of *flos* to *menstrua*. Yet he also added the following explanation for why the *menstrua* are colloquially called 'woman's flower': 'the menses … are commonly called "the flowers" among women, for just as trees without flowers do not bear fruit, likewise women without their flowers are deprived of the function of conception' (Anon., *Conditions of Women* f. 247r). Even though the author was probably influenced by a literary flourish in his Arabic source (where the menses were likened to the liquid excrescences of trees), the use of the analogy of 'flowers' must have come, as he himself acknowledges, from the word women themselves commonly used. This is reflected in the fact that, when *Conditions of Women* came to be translated into a variety of different vernacular European languages, the term most often chosen to render the Latin *menstrua* was 'flowers' (Green, *The 'Trotula'* 21). This was true across the board – from the Romance languages to Germanic ones and even Hebrew. Italian had *i fiori*, Catalan *la flor*, Dutch and German *bloemen* and *Blumen*, respectively. If 'the flowers' was an invented usage, then it was one that had popular appeal. Formal medical and ecclesiastical Latin continued to prefer *menstrua* (or the singular, *menstruum*) and vernacular medical texts might also use other terms such as the Dutch *stonden* ('periods'), German *dy suberunge* ('purgation'), and Italian *ragione* ('the rule'). Importantly, none of these other terms were inherently negative nor were they necessarily euphemistic, as if there was some other term that was deemed too explicit to be used in public discourse.

Anatomy and physiology

Menstruation formed the bedrock of medieval concepts about how the female body functioned. Our modern understanding of menstruation sees it as the sloughing off of the endometrial tissue when no fertilised ovum has been implanted in the uterine wall; in other words, menstruation is a process confined to the uterus, the function of which is strictly reproductive.[3] In medieval Europe, menstruation was seen as the end result of a whole bodily process of purification, one unique to the female body. Menstruation was also seen, as the quotation from the *Conditions of Women* above shows, as a necessary prerequisite for conception. In fact, many references to menstruation in the context of fertility see the successful completion of its purgative function as key to fertility: the womb must be properly cleaned out before it is able to provide a suitable 'field' for the seed. This concept is, of course, crucial to understanding why emmenagogues (preparations for bringing on menstruation) were not necessarily seen as abortifacients but may well have had a deliberately pronatal purpose (Green, *The 'Trotula'* 182, n. 144). Although the uterus was seen universally as the principal organ of menstruation, the anatomical basis of the menstrual function was never explored in any depth.

Constantine the African, for example, who translated a variety of Arabic medical texts into Latin, never mentions menstruation when he describes the functions of the uterus. A few decades later, the Salernitan writer Copho, in his *Anatomy of the Pig*, would articulate the common medieval view: '[i]t is known that Nature arranged this organ in women so that whatever superfluity of the whole body is generated in them might be sent out through this organ at the accustomed time, just like bilge-water, and hence women naturally have menstruation' (144). In medical writings, there is occasional reference to the functions of the veins in the uterus, as if their narrowness or openness controlled the blood that flowed to the uterus from all parts of the body. Thus, in the Salernitan text *Conditions of Women*, it is said that very emaciated women can suffer menstrual retention because the veins in their womb are too small to allow the superfluous humours to erupt. More commonly, however, menstrual difficulties were ascribed to the bodily humours themselves (being too thick or too thin) rather than to any anatomical defect.

Precisely because menstruation was seen as a physiological function of the *whole* female body, its disruption (particularly its retention) was seen as the cause (and not as merely the symptom) of a whole range of other disorders, from skin conditions to breast cancer, heart palpitations, suffocation, and even death. Breast cancer was linked to menstrual retention because, it was thought, special veins linked the uterus to the breasts so that after pregnancy the blood that had nourished the foetus *in utero* could be redirected up to the breasts and there converted into milk. In middle-aged women who had ceased to menstruate, the unexpelled corrupt blood migrated naturally to the breasts, there producing lesions. Given the severity of illness that disruption in the normal menstrual processes could produce, it is little wonder that menstrual conditions (especially retention) were the principal condition against which therapeutic interventions in women would be offered. For example, of some 250 medicinal substances listed in the most popular pharmaceutical authority, the *Circa Instans*, eighty-seven – over a third – are stated to be good for menstrual disorders.

Because it was expected to be periodic, 'normalcy' of menstruation was defined not simply by quantity, consistency, or colour, but also by its timing. It should come once every thirty days; the duration should vary from two days in some women to four in others. The age of onset is usually given as around fourteen, while forty or fifty is often cited as the common age of cessation.[4] These ages were inherited from classical and Arabic medical writings and generally cannot be relied on as demographically accurate. Nevertheless, there are some interesting cases where it seems that translators or scribes are offering their own observations of the normative ranges among women in their area. A cleric writing in London in the latter part of the fourteenth century observed a recent drop in the age of menarche: 'In ancient times, the menses did not begin to flow until the fifteenth or fourteenth year, or certainly not before age twelve. But now they begin in certain girls in the eleventh or in the tenth year. And at that point they are capable of conception' (Mirfeld, f. 122va–b). This remarkable comment is all the more intriguing as it comes *after* the devastations of the Black Death in the mid-fourteenth century. A fifteenth-century Dutch translator of the *Conditions of Women* gave the menarcheal age-range as between nine and sixteen. A Latin scribe writing in the Low Countries a few years later, however, originally wrote 'eleven' as the age of menarche, but then crossed it out and wrote 'eighteen' (Green, *The 'Trotula'* 175–6). A text from the early Middle Ages entitled *The Epistle on Virgins* suggests that a woman's later fecundity and health were determined by the age at which menstruation begins. Thus, if a girl began to menstruate at age ten, her menses would last until she is thirty (or twenty, in one version) and, although she would live a long life, she would be afflicted by many pains and bear only one child. Fifteen was the best age at which to begin to menstruate, as it augured well for high fecundity, low morbidity, and long life (Fischer 121).

In the later Middle Ages, this rather deterministic view of menarche disappeared. Instead, the only age differential was that adolescent women were believed to menstruate in the first phase of the moon, women in the prime of life in the second phase, and older women in the third or fourth phase. The age ranges mentioned for menopause varied even more widely. The Latin *Conditions of Women* placed the average age of cessation of menstruation at fifty, with thin women menstruating as late as sixty-five and fat women stopping as early as thirty-five (it may have been that fat women were assumed to convert their excesses to fat rather than menses.) Hildegard of Bingen, also writing in the twelfth century, believed that women could continue to menstruate up to age eighty (146–7).[5] A woman of early fourteenth-century France, Beatrice de Planissoles, in her testimony before some Inquisitors, notes with surprise how fervid her desire was toward one of her lovers, even though she had already passed menopause (qtd. Fournier 200). Clearly, she had some sense that menstruation (and not simply age alone) was connected to sexual pleasure.

Complete absence of menstruation in a woman of child-bearing age was generally seen as pathological. A second-century Greek writer, Soranus of Ephesus, had hypothesised that certain women – athletes, dancers, and singers – could remain healthy even without menstruating since their active lifestyles were in perfect balance with the amounts of foods they took in. Soranus's views on the possibility of salubrious amenorrhea passed to the early Latin West when his works were translated in late Antiquity, but then were abandoned with the new influence of the Galenic theory of the humours in the twelfth century. Nevertheless, there was one interesting survival of the Soranic view. Whereas some medical writers believed that nuns (and often rich women generally) needed to menstruate *more* than other women because of their rich diets, an English writer in the late fourteenth century instead suggested that 'because of rising at night and singing and occupation in their [religious] service,' nuns' blood is likely to waste away more than other women's (Anon., 'Nature of Women' 86). The particular physiology of nuns may well have been an issue of speculation, since most religious rules stipulated that phlebotomy (blood-letting) be performed regularly four times a year. It was duly noted that Joan of Arc had ceased to menstruate when she was being held captive by the English forces (McCracken 31).

Menstruation could also be excessive while pathological fluxes (variously described by their colour or humoural character) were recognised. From the fourteenth century on, medical texts often include a chapter on 'the white flowers' or 'white menses', a vaginal discharge which may have been lochial flow. There is little evidence, however, that much thought was given to differences between menstrual blood and the lochial flow (usually called simply the 'flux of blood'). Although these two kinds of uterine bleeding were distinguished terminologically, few commentators gave thought to what, precisely, distinguished these two flows besides their timing and duration. An exception is a mid-fifteenth-century writer, Pierre Andrieu, who suggested that the lochial flow had a 'more vehement blackness' than menstrual blood because it had been retained longer in the womb (f. 89ra). Complete absence of the lochial flow was recognised as a pathological condition and medical writers recommended that, if it did not appear on its own, it needed to be induced. Visual depictions of menstrual blood in medical contexts are virtually non-existent. Unlike uroscopy charts that depicted, often in quite vivid colours, the different appearances of the urine, examination of menstrual blood seems to have been too rare to elicit iconographic depiction. A rare exception is a thirteenth-century manuscript of the pharmaceutical authority, Dioscorides, which includes elaborate illustrations of various medical practices described in the text. In scenes that show treatments for the flux of blood or *fluxus sanguinis* (ff. 69r, 89r, and 115r), the genitalia are not depicted.[6] Rather, the iconographic signal is simply to show the woman with her skirts raised, in some cases above her knees, with small drops of blood depicted falling between her legs.

A striking feature of medical writing is that it keeps a largely neutral perspective on menstrual blood. It is very rare for a medical writer to say anything about the allegedly harmful properties of menstrual blood. Just as other bodily emissions, such as urine, might be usefully examined for signs of internal disruption in the body, so might doctors ask to see a woman's menstrual cloth to determine which humour might be predominating in her body. In the early Middle Ages, menstrual blood was actually recommended as a medicinal substance. This practice seems to have died out by the twelfth century (as did most other uses of human effluvia), but certain magical practices had a longer life. Burchard of Worms, a German bishop of the early eleventh century, identified what he believed was a popular superstition among women: that is, that they take their menstrual blood and give it, mixed in with food or drink, to their husbands so that they will be loved more by them (Burchard, bk. XIX, ch. 5, 974C). The accused heretic Beatrice de Planissoles stated that she had kept the first menstrual rag of her daughter in order to perform love magic. There are, however, a few exceptions to this largely neutral medical view of menstruation. Sexual intercourse during menstruation was commonly seen as a cause of leprosy or deformity in any conceived offspring. The fourteenth-century English surgical writer, John Arderne, in giving advice on decorum to the aspiring (male) practitioner, says that he should avoid substances or behaviours that might make his breath annoying to his patient; included among the latter is having recently had sex with a woman who is menstruating (f. 88, 1.11). He also contends that the breath of a menstruating woman will interfere with the healing of wounds if she draws near (f. 88, 1.9).

Menstruation in medieval science and religion

In modern western cultures, a distinction between scientific and medical views of menstruation would probably be meaningless, so pervasively has biomedicine grounded itself on basic scientific principles of anatomy, endocrinology, and so on. In the Middle Ages, however, such a distinction still had some meaning, since the major scientific authority, Aristotle, had differing views on menstrual blood and generation than did the chief medical authority, Galen. In medical texts, as we have seen, the stress is on menstruation as a necessary purgation. Medical texts do not usually engage in speculation about the development of the foetus, but it seems to be understood that any new 'menstrual' blood that forms during the pregnancy goes to nourish the foetus. Equally important, in strictly Galenic medical views, was the understanding of what menstruation was *not*; it was not the 'seminal contribution' of the female to conception. Rather, in Galen's view, women (who had 'testicles' just like men did) produced their own (albeit weaker) seminal fluid which was 'ejaculated' into the uterus upon sexual climax. Aristotle, in contrast, had been writing before the ancient Alexandrian anatomists had discovered the 'female testicles' and asserted their functional similarity to the male ones. For Aristotle, the menstrual blood itself was the sole maternal contribution to reproduction: it was the 'matter' on which the 'form' of the new embryo was imprinted by the male seed.

For medieval writers, who inherited both the Aristotelian line of thinking and the Galenic one, the subject of the role of the menses was therefore unclear. Constantine the African himself demonstrates this confusion. In his translation of al-Majusi's Arabic *The Whole Art of Medicine*, Constantine seems initially to be giving an Aristotelian equation of male semen with female menses, explaining that '[t]he semen is both the creative essence and the matter of the infant-to-be, while the menses are only the matter'. Yet he immediately goes on to state that *both* parents contribute seed. 'No foetus is created except when the seeds of the male and the female are mixed' (315). This confusion or conflict between the Galenic and Aristotelian positions can give some sense why a text like the c.1300 *Secrets of Women*, which draws on both natural-philosophical and medical traditions, should both be so obsessed with explaining the menses and so concerned about what effect they might have both on the foetus and on other people.[7]

The *Secrets of Women* also picks up on another tradition, one from which the medical tradition had largely been insulated. This can be called the Plinian tradition, although the first-century Roman naturalist was often merely relaying lore he gathered from the culture around him. Pliny created the standard litany of the menses' ill effects (they can kill crops, rust iron, and so on) but he also itemised a variety of medical uses for them (bk. 28, sect. 23). The negative moreso than the positive views were later echoed in such late antique and early medieval writers as the third-century Solinus and the sixth/seventh-century encyclopaedist Isidore of Seville (d. 636), whence they passed into the learned lore of countless medieval priests and other literates. It is difficult to say how much these views were shared among the broader populace. However, taboos surrounding menstruation seem to have been widespread. It is striking, nonetheless, how little effect the Plinian tradition had in medical writing. No suggestion of the idea that the menses were poisonous is found in any version of the *Trotula* texts, Latin or vernacular, until the very end of the Middle Ages. Interestingly, this one exceptional text, an English translation of *Conditions of Women*, while warning of the dangers of sex during one's period, nevertheless excuses any woman from blame who 'cannot by any excuse or entreaty defend [against] her husband' (Anon., *Book called Trotela* f. 33r). A rare moment when medical science did pursue Plinian concepts vigorously was a late medieval episode, largely confined to Spain, when physicians attempted to explain how it was physiologically possible for a postmenopausal woman to inflict the evil eye on children and other unwary victims (Salmon and Cabré *passim*).

Medieval Islam, Judaism, and Christianity were united in their belief that a woman was ritually unclean during menstruation and that therefore intercourse with her should be avoided at this time. The three monotheistic religions connected, to varying degrees, the origin of menstruation with Eve's sin in Paradise (Spellberg *passim*). On other aspects of menstruation, attitudes and practices diverged. We know virtually nothing about the specific menstrual practices of the Muslim women who lived in medieval Europe. For Jewish women, there is ample evidence of practices surrounding use of the *mikveh*, the ritual bath, and the social interactions that surrounded it.[8] Observance of menstrual regulations was one of the most important duties of a Jewish woman; failure could, among other things, be the cause of death in childbirth (Baumgarten 40). Indeed, it has been suggested that Jewish notions about the polluting effects of menstruating women were so strong as to make medieval Judaism the only religious tradition that did not produce any female mystics (Koren *passim*). Western Christian tradition developed no comparable menstrual rituals, though ideas of women's ritual uncleanliness did affect religious practice.[9] Early Christian practices forbade menstruating or postpartum women from entering church or taking communion. In the late sixth century, Pope Gregory the Great challenged this exclusion, arguing that since menstruation was an involuntary 'infirmity', menstruating (or postpartal) women should not be excluded from church. Gregory's views were not universally endorsed, and exclusionary practices continued into the twelfth century when Pope Innocent III put an end to the controversy: although a menstruating (or postpartal) woman should not be discouraged if she *wanted* to stay out of church, she could not be prohibited. The Levitical injunctions were part of the 'old law' that Christ's coming had abolished (Miramon *passim*). Changes in attitudes towards women's ritual impurity during menstruation or the postpartum period did not change negative attitudes toward the blood itself. Church law continued to prohibit sexual intercourse during menstruation, with some canonists fearing that violators would beget lepers and monsters (Brundage 91–92, 156, 199, 242, 283, 451, 508). The practice of lying-in after birth (usually for about a month) and the ceremony of purification or 'churching' to reintegrate the woman into the community of the faithful likewise included an element of 'cleansing' from a perceived impurity as well as thanksgiving for survival through the ordeal of childbirth (Rieder *passim*). So ingrained was this idea that menstruation was inherently defiling that theologians debated whether Mary, the mother of Jesus Christ, had menstruated and so been afflicted by 'Eve's curse' in the same way other women were (Wood *passim*).

Male menstruation

Perhaps the most intriguing medieval notion about menstruation – intriguing both because it strikes the modern mind as so unlikely and because it endured so long – is the notion that men, particularly Jewish men, might also menstruate. This notion seems to have arisen out of an idea, common among Salernitan medical writers, that bleeding haemorrhoids in men might actually be salubrious and, as such, should not be interfered with medicinally. In fact, haemorrhoidal bleeding might even be actively provoked in the same way menstrual bleeding was, and for the same reason: to release excess corrupted humours that could not otherwise break free. Thus we find, for example, recipes said to be good for provoking both the menses and haemorrhoidal bleeding. The late twelfth-century writer, Bernard of Provence, who studied at Salerno, listed over fifty different substances to provoke both the menses and haemorrhoids, making clear in the case of suppositories that whereas women were to insert theirs into the vagina, men should insert theirs into the anus (Bernard 282–4). A few years later, Roger de Barone describes the equivalence as follows:

> This flux is sometimes natural, sometimes unnatural, just as the menstrual flux is in women. Hence, just as women menstruate each month, so some men suffer from the haemorrhoidal flux each month, some four times a year, some once a year. This flux ought not be restrained, because it cleans the body of many superfluities. (qtd. Green, *The 'Trotula'* 175)

There is nothing here, of course, to suggest that any particular kind of male is more subject to this bleeding than another. It is nevertheless striking how explicitly Roger draws the parallel between men and women, particularly in stressing the periodicity of this bleeding.

Within about a hundred years of Roger, ideas about periodic male bleeding took a different turn. Peter Biller has noticed a striking coincidence in the late thirteenth and early fourteenth century: at the same time Aristotle's natural philosophical writings were being debated among Parisian theologians, there was also a series of expulsions of Jews from Gascony, England and France by kings and other lords fearing that they could not contain rising anti-Semitic sentiments among the majority Christian population. Biller identifies three separate strands of thought, all deriving from the Arabic world, that came together in this period: medical texts claimed that melancholy was associated with a flux of blood; astrological texts claimed that Jews were associated with melancholy and the morbid influences of the planet Saturn; and two texts coming out of the Crusader states claimed that Jews suffered a flux of blood as punishment for their role in the death of Christ. The more flagrant statements on Jewish male menstruation became both anti-Semitic and, at their root, profoundly misogynistic. While other chapters in this volume explore this notion more fully, here it is important to note that medical notions of humoural flux and excess were, at least at an early point in the development of ideas on male haemorrhoidal bleeding, strong enough to allow an analogy between the male and female body that did not automatically heap opprobrium on the latter.

Enclosures

This chapter has focused most on medical discussions of menstruation because that is where the majority of explicit references to this bodily process can be found. Much harder for the historian to gauge is the meaning of the *silence* that surrounds this topic in so many other areas. In Middle English, aside from a handful of uses in Biblical texts (where menstruation always carries a notion of uncleanliness) and alchemical texts (where *menstrues* is used neutrally to refer to a precipitate of a mineral or chemical), the words *flowers* (in the sense of 'menses') and *menstrues*, and its derivative forms, have been found only in medical texts. As Peggy McCracken has noted in her survey of attitudes toward blood in medieval French literature, outside of medical contexts menstruation is largely seen as *nefas*, something that cannot or should not be named or spoken about. Thus, even though such genres as the *fabliaux* could engage in salacious repartee about the sexual organs and sexual functions, menstruation is never mentioned. This is particularly intriguing since, one might imagine, violation of religious injunctions against sex during menstruation might have had its own comic possibilities. The fact that medieval people seemed unwilling even to joke about menstruation (at least in mixed company) may be our most important clue in deciphering common attitudes about this bodily function.

In an intriguingly candid passage, a later medieval French translator of the *Trotula* prefaced his text with some advice to his fellow male practitioners. Moving beyond the standard topos of women's shame preventing them from baring their ills to male physicians, he offered his frustrated fellow practitioners some additional aid:

> Because women are ashamed to name these things [the reproductive organs and their functions] in the manner in which we are accustomed to name them in French, we will name each organ by another name which they do not customarily have. Thus, women will name them more readily without being ashamed. (Camus 109)

He then provided a list of these terms, among which he includes *les fleurs* as a substitute for the 'customary' name *les maladies secrettes*. That even he treats the deliberately opaque euphemism *les maladies secrettes* as the normative French term for the menses says a great deal about the increasing layers of verbal and discursive enclosure of women's bodies and bodily processes in the late Middle Ages.[10]

Notes

1. All translations, unless otherwise noted, are my own.
2. The vast majority of medieval gynaecological texts have never been published. For a comprehensive listing of all known medieval works on women's medicine see my *Women's Healthcare*.
3. There is no universal agreement even amongst today's scientists of the functions of menstruation. Theories currently in circulation include the notion that menstruation serves to flush out sperm and other micro-organisms that may enter the body through the vagina (Clough *passim*; Finn *passim*).
4. There are no specific terms for either 'menarche' or 'menopause' in medieval medical writing, nor is either seen as a specific medical event that is problematised as needing 'treatment'.
5. Moulinier convincingly demonstrates that the *Causes and Cures*, as it now exists, was compiled after Hildegard's death from bits and pieces of her authentic writings. Even if these views on menopause are not literally from Hildegard's hand, they were plausible to her immediate peers.
6. This is something which I have found to be true universally in medieval medical illustrations for females, in drastic contrast to those for males.
7. See Bildhauer, this collection, for more on the *Secrets of Women*.

8. See, for example, Marienberg *passim*.
9. See Luke 8:43–8, Mark 5:29–34, and Matthew 9:20–2.
10. See my 'From "Diseases of Women" ' *passim*.

Works cited

Andrieu, Pierre. *Pomum Aureum*. MS lat. 6992, s. xv med., ff. 79r–90v. Bibliothèque Nationale de France, Paris.

Anon. *Book called Trotela*. MS Longleat 333, s. xvi2, ff. 33r–43v. Longleat House, Warminster.

Anon. *Liber de sinthomatibus mulierum [Conditions of Women]*. MS 173, ff. 246v–253r. Magdalen College Library, Oxford.

Anon. 'Nature of Women'. *Studies in the Age of Chaucer* 14 (1992): 53–88.

Anon. *Secres de Femmes*. MS O.1.20 (1044), s. xiii med., ff. 21rb–23rb. Trinity College Library, Cambridge.

Arderne, John. *Fistula in Ano*. c. 1425. MS Sloane 6. British Library, London.

Baumgarten, Elisheva. *Mothers and Children: Jewish Family Life in Medieval Europe*. Princeton: Princeton University Press, 2004.

Bernard of Provence. *Commentarium Magistri Bernardi Provincialis super Tabulas Salerni*. *Collectio Salernitana ossia documenti inediti, e trattati di medicina appartenenti alla scuola medica salernitana*. 1852–59. 5 vols. Ed. Salvatore De Renzi. Bologna: Forni, 1967. Vol. 5. 269–328.

Biller, Peter. 'A "Scientific" View of Jews from Paris around 1300'. *Micrologus* 9 (2001): 137–68.

Brundage, James A. *Law, Sex, and Christian Society in Medieval Europe*. Chicago: Chicago University Press, 1987.

Burchard of Worms. *Decretorum libri viginti*. *Patrologia Latina*. 1844–45. Ed. Jacques-Paul Migne. Alexandria: Chadwyck-Healey, 1995. Vol. 140.

Caelius Aurelianus. *Caelius Aurelianus, Gynaecia: Fragments of a Latin Version of Soranus' Gynaecia from a Thirteenth Century Manuscript*. Ed. Miriam and Israel Drabkin. Baltimore: Johns Hopkins University Press, 1951.

Camus, Jules. 'La Seconde Traduction de la Chirurgie de Mondeville (Turin, Bibl. nat. L.IV.17)'. *Bulletin de la Société des Anciens Textes Français* 28 (1902): 100–19.

Clough, Sharyn. 'What is Menstruation For? On the Projectibility of Functional Predicates in Menstruation Research'. *Studies in History and Philosophy of Biological and Biomedical Sciences* 33.4 (Dec. 2002): 719–32.

Constantine the African. '*De genitalibus membris*. The *De genecia* attributed to Constantine the African'. Ed. Monica. H. Green. *Healthcare in the Medieval West: Texts and Contexts*. Aldershot: Ashgate, 2000. 312–23.

Copho. *Anatomia porci*. 'Die erste Tieranatomie von Salerno und ein neuer salernitanischer Anatomietext'. *Archiv für Geschichte der Mathematik, der Naturwissenschaften, und der Technik* 10 (1927): 136–54.

Dioscorides. *Medicina antiqua*. Trans. Reinhild Weiss. London: Harvey Miller, 1999.

Finn, Colin A. 'Why do Women Menstruate? Historical and Evolutionary Review'. *European Journal of Obstetrics, Gynaecology and Reproductive Biology* 70.1 (Dec. 1996): 3–8.

Fischer, Klaus-Dietrich. 'Die pseudohippokratische *Epistula de virginibus*: Bemerkungen zu ihrer Textüberlieferung und zu ihrem Vokabular'. *Les Études Classiques* 70 (2002): 101–22.

'Flower'. *Oxford English Dictionary*. 2nd ed. 1989.

Fournier, Jacques. 'Inquisition Records'. *The Later Middle Ages*. Vol. 2 of *Readings in Medieval History*. 2 vols. Ed. Patrick J. Geary. Ontario: Broadview, 1996. 185–204.

Green, Monica H. 'From "Diseases of Women" to "Secrets of Women": The Transformation of Gynecological Literature in the Later Middle Ages'. *Journal of Medieval and Early Modern Studies* 30 (2000): 5–39.

Green, Monica H., ed. and trans. *The 'Trotula': An English Translation of the Medieval Compendium of Women's Medicine*. Philadelphia: Pennsylvania University Press, 2002.

Green, Monica H., ed. *Women's Healthcare in the Medieval West: Texts and Contexts*. Aldershot: Ashgate, 2000.

Hildegard of Bingen. *'Beate Hildegardis Cause et cure'. Rarissima Mediaevalia*. Ed. Laurence Moulinier. Berlin: Akademie Verlag, 2003.

Hunt, Tony. *Anglo-Norman Medicine*. 2 vols. Cambridge: D. S. Brewer, 1994–97.

Koren, Sharon Faye. 'The Woman from Whom God Wanders: The Menstruant in Medieval Jewish Mysticism'. PhD Diss. Yale University, 1999.

Marienberg, Evyatar. 'Niddah: Etudes sur la "Baraita de Niddah" et sur la conceptualisation de la menstruation dans le monde juif et son écho dans le monde chrétien de l'époque médiéval à nos jours'. PhD Diss. École des Hautes Études en Sciences Sociales, 2002.

McCracken, Peggy. *The Curse of Eve, the Wound of the Hero: Blood, Gender, and Medieval Literature*. Philadelphia: Pennsylvania University Press, 2003.

Miramon, Charles de. 'Déconstruction et reconstruction du tabou de la femme menstruée (XII–XIIIe siècle)'. *Kontinuitäten und Zäsuren in der Europäischen Rechtsgeschichte: Europäisches Forum Junger Rechtshistorikerinnen und Rechshistoriker, München 22–24 Juli 1998*. Ed. Andreas Their, Guido Pfeifer and Philipp Grzimek. Frankfurt: Peter Lang, 1999. 79–107.

Mirfeld, John. *Breviarium Bartholomei*. Harley MS 3, s. xiv ex. British Library, London.

Moulinier, Laurence. 'Le corps des jeunes filles dans les traités médicaux du Moyen Âge'. *Le corps des jeunes filles de l'Antiquité à nos jours*. Ed. L. Bruit Zaidman *et al*. Paris: Perrin, 2001. 80–109.

Pliny the Elder. *Historia Naturalis*. Berlin: Teubner, 1897.

Rieder, Paula. 'Between the Pure and the Polluted: The Churching of Women in Medieval Northern France, 1100–1500'. PhD Diss. University of Illinois at Urbana-Champaign, 2000.

Salmon, Fernando and Montserrat Cabré i Pairet. 'Fascinating Women: The Evil Eye in Medical Scholasticism'. *Medicine from the Black Death to the Great Pox*. Ed. Jon Arrizabalaga *et al*. Aldershot: Ashgate, 1998. 53–84.

Spellberg, D. A. 'Writing the Unwritten Life of the Islamic Eve: Menstruation and the Demonization of Motherhood'. *International Journal of Middle East Studies* 28.3 (Aug. 1996): 305–24.

Wood, Charles T. 'The Doctors' Dilemma: Sin, Salvation, and the Menstrual Cycle in Medieval Thought'. *Speculum: A Journal of Medieval Studies* 56.4 (1981): 701–27.

5

The *Secrets of Women* (c. 1300): A Medieval Perspective on Menstruation

Bettina Bildhauer

If menstruation has always been shrouded in mystery and taboo, this is nowhere more the case than in the very research that makes such claims. Studying menstruation still has an air of being – excitingly or inappropriately – taboo-breaking; menstrual beliefs of other cultures and times are still often presented as mysterious, exotic idiosyncrasies, or as furnishing hidden proof for the fundamentally identical preoccupations of all cultures. That such mystification of menstruation in scholarly discourse has a long and inglorious history is apparent, for instance, in the medieval treatise *Secreta Mulierum (Secrets of Women)*. These 'secrets of women' primarily concern menstruation and its pivotal role in reproduction – they are secrets *of* women both in the sense of pertaining to women, and of being kept secret from men by women. But the text then shares precisely this arcane knowledge within a community of men and combines it with common misogynistic stereotypes about the wickedness and uncontrollability of women. The rhetoric that declares menstruation secret is used to create a gender distinction that is precisely not mysterious or universal, but has the historically specific context and function of excluding women from scholarly discourse. In ways which I shall now explore, the *Secreta* uses menstruation to give scientific confirmation of a view of women as opposing and polluting men, as precariously enveloping them and as sucking them dry – a fragile gender system based on, but also destabilised by, menstruation.

The *Secreta Mulierum*, according to Monica Green, 'is beginning to be rightly acknowledged as one of the most influential documents in the history of medieval scientific attitudes toward women' (14–15).[1] Wrongly attributed to Albert the Great throughout the Middle Ages, it was most probably written by a German cleric in the late thirteenth century. Although menstruation was discussed in many different contexts in the Middle Ages, such as theology and courtly literature, this popular treatise (with about one hundred extant manuscripts) is one of the most extensive works on the topic. The *Secreta* presents standard views on menstruation that have been frequently expressed in learned writings since antiquity and were well-known among the literate elite, but it also evokes them in a particularly exhaustive and misogynistic way.[2] The text is concerned with natural philosophy rather than medical practice, and with topics such as sexual intercourse, conception, development of the embryo and foetus, childbirth, pregnancy, virginity, infertility, and male semen. It was accompanied by commentaries from very early on, and, while originally written in Latin, was subsequently translated into German and other languages. I shall here use an anonymous Southern German translation dating from around the third quarter of the fifteenth century, probably produced for an urban audience.[3]

Menstruation and pollution

One of the *Secreta*'s learned opinions that might seem particularly mystifying to modern Western readers is the belief that menstruating women stain mirrors: 'If a woman has this [menstrual] flow and looks into a mirror during this time, this mirror becomes like a bloody cloud. And if the mirror is new, one can hardly remove the red staining from the mirror, but if it is old, one can easily remove it' (626–32). This idea was not a one-off curiosity, but a staple of scholarly writing since Pliny and Aristotle.[4] If we are to understand how such a view made sense to ancient and medieval audiences, it is worth noting first of all that this statement stresses the polluting effect of menstruation: the woman's gaze is described as something dirtying the clear mirror, making it so cloudy that it has to be subsequently cleaned. The defiling nature of menstrual fluid is highlighted in a number of further statements in the *Secreta* as well. Menstruation's relation to the rest of the body is compared to that of excrement in a public toilet: 'And therefore Avicenna says that the uterus in women is like a toilet that stands in the middle of town and to which people go to defecate, just like all residues of the blood from all over the woman's body go to the uterus and are cleaned there' (2416–19). It is also referred to as a 'disease' or even 'cancer' (2363), as analogous with venomous snakes (980–9; 1068–72) and as poisonous to a woman if retained in her body (2390–430).

Mary Douglas's compelling analysis of pollution famously uses Lord Palmerston's old adage of dirt as 'matter out of place' to highlight that polluting substances have to be seen in a wider context: menstrual blood is not polluting or dirty in itself, it is only so if it diverges from or undermines a system (36). This system, according to Douglas, is the social order that is symbolised by the individual body:

> The body can stand for any bounded system. Its boundaries can represent any boundaries which are threatened or precarious. The body is a complex structure. The function of its different parts and their relation afford a source of symbols for other complex structures. We cannot possibly interpret rituals concerning excreta, breast milk, saliva and the rest unless we are prepared to see in the body a symbol of society, and to see the powers and dangers credited to social structure reproduced in small on the human body. (116)

Menstrual blood in the *Secreta* is presented as upsetting the wider cosmic balance, especially of gender relations. But, unlike Douglas, I would argue that this is not just the case because the body symbolises the social order of gender, but because gender is a bodily phenomenon in itself. Menstruation disrupts not just gender roles, but also the integrity of individual bodies, which is as important and threatening as its undermining of more abstract structures. Moreover, for all its 'polluting' potential, menstruation is a vital commonplace disruption suggesting an understanding of bodies as bounded entities, bodies the delimiting of which also establishes maleness and femaleness as distinct.

In accordance with ancient Greek philosophy and science, the *Secreta* considers the macrocosm of the universe and the microcosm of the body as corresponding and connecting with one another. The primary means of communication between all elements of the cosmos were thought to be the two basic qualities of temperature and moisture and the combinations of hot or cold, and dry or moist. Human bodies were perceived to be essentially containers of the four bodily fluids or 'humours', which each represented a particular combination of temperature and moisture: blood, black bile, yellow bile and phlegm. Their balance was responsible for the 'complexion', the temperament and also the personality and health of the person. Blood was seen as the most important humour, which transported all three others. All bodily fluids, including not only the other humours, but also menstrual blood, semen and breast-milk, were processed forms or by-products of blood, which in turn was digested food. In this intricately linked cosmos, where an equilibrium of heat and moisture is of utmost importance, woman is presented as a human who leaks. Part of her food, rather than being fully digested, simply seeps out, in the form of menstrual blood. This means that both the cosmic balance and the boundedness of her body are disrupted by menstruation. However, this uncontrolled menstrual flow is precisely what allows a stable distinction between the genders: it is what makes woman different from man throughout the text, especially in her main role as sexual partner and mother. Menstrual blood, for example, supposedly incited women's sexual desire and rapacious appetite for extracting men's fluids. It is the woman's contribution to the embryo in conception, it nourishes the foetus in the uterus and, having been processed into breastmilk, provides food for the newborn, too. In this way, women are defined by menstrual blood in all of the aspects with which this treatise is concerned – sexuality, conception and reproduction.

Menstruation and gendered binaries

Menstruation in the *Secreta* is understood as a category of universal significance through equation with a range of further concepts in addition to femaleness, including matter, the uterus, cold, the left-hand side and the outside, oral communication, the body and Jewishness. Masculinity, on the other hand, defines itself in opposition to this category – as coherent, as able to exchange written knowledge about women's bodies, and as associated with intellect, form, warmth, the right-hand side, the inside and Christianity. For instance, Johann Hartlieb's fifteenth-century translation of the *Secreta* uses menstruation to map a gender hierarchy onto the distinction between Jews and gentiles, by claiming that Jewish men also menstruate: 'all Jews are used to having the [menstrual] flow every month' (Bosselmann-Cyran, p. 136, Ch. 15, ll. 44f.).[5] Moreover, the Aristotelian alignment of women with base matter, substance or body, and of men with superior form, idea or spirit is extended in various directions in the *Secreta*. The treatise cites both of the main theories about conception circulating in the Middle Ages (111–21): the Galenic two-seed theory that both a male and a female seed contribute to the essence of the embryo and the Aristotelian theory that the male seed imprints the form of the embryo on the raw material, the matter, of maternal menstrual blood.[6] While ostensibly leaving the choice of which theory to believe in to the reader, the text goes on to conflate the two, and never clearly distinguishes between menstrual fluid and female seed. Ultimately, however, the idea of a female seed is submerged, since it is usually simply referred to as *menstrum* or matter *(materj)* – the concepts used in Aristotelian theory. The connotations of form and matter are expanded in the image of the penis as a farmer sowing a field with his male seed, an image which aligns femininity with nature being cultivated by men (2546–51). The distinction between matter and form is also applied on the level of the text itself: women provide the subject matter of the *Secreta*, which men bring into a written form: the author is referred to as a male 'master', who writes to a 'fellowship' of men and invites further commentaries and contributions from them. If women are granted any knowledge at all, it pertains to their bodies: women know when they are about to give birth, how to feign virginity; how to induce a miscarriage, how to conceal pregnancy, and how to wound the penis in the vagina (1692–4; 2299–2309, 2115f., 2357–60; 1405–12, 1448–55; 2110–28, 2216–35; 584–625).

For all its attempts at distancing itself from women's matter, however, the treatise itself mainly relates knowledge about the body, that is, exactly the kind of knowledge it has vilified as dangerous and female intelligence. The 'secrets of women', their sexual functions that have to be modestly hidden, become the *Secreta Mulierum*, a body of male knowledge that is equally esoteric and has to be concealed from susceptible minds, as the prologue advises. Similarly, the word 'matter' or *matery* is used to refer both to menstrual matter and to the subject matter of the treatise, and can be employed ambiguously, for instance, in the first chapter, where menstruation *is* the topic. So the learned male discourse is linguistically 'contaminated' by the menstrual matter.[7] The role of men in the text is also unstable because knowledge about the male body is entirely dependent on and secondary to that of the female body. This is visible even in the structuring of the treatise: the first chapter, concerning the physiology of menstruation, is followed by a discussion about why superfluous food constantly seeps only out of women and not out of men, whose emission of semen is much more sporadic and controlled. This is explained by men's warmer nature, which allows them to digest food more effectively, thus leaving fewer poisonous residues that need to be expelled. So menstruation – leakage – is the given, and male identity is defined in contrast to it, as that which does *not* leak. The structure of the *Secreta* as a whole with its one, final, chapter on semen shows the same treatment of maleness as an addendum to its main concern with female physiology.

If semen is the male equivalent of menstrual blood, every attempt is made to distinguish between the two. Two different concepts of fluidity emerge: on the one hand, women's negative fluids, which threaten stability; and on the other hand, men's fluids, which function as the source of life.[8] One of the main differences between these two substances is that women's fluidity is uncontrollable, whereas the male kind is more coherent and almost solid. If breastmilk, for example, is well-digested, thick and well-formed, this has nothing to do with the woman's qualities, but is a sign of a male foetus warming his mother's body sufficiently to allow better processing: 'And if thick and well-formed milk flows and runs from her breasts – milk that is so thick that if one took this milk and poured it on a board or another flat surface, it would stay together –, that is another sign that the offspring is male' (2173–6). This is contrasted with the thin, undigested, dissolving, uncontrollable milk produced under the influence of the female foetus, which flows apart like water on a board (2186–9). The same appreciation of thickness occurs in the description of semen itself, which, if coagulated like boiled milk or pitch, produces strong – and of course, more often male – children, whereas thin, liquid semen brings forth fragile fruit (2944–69). But again, these attempts at distinguishing male from female fluids betray nothing so much as their similarity.

Menstruation and its dangers for men

While any women's bleeding potentially upsets the cosmic balance of fluids, the immediate victims are men. Pregnant women are imagined as leaking containers. For example, in a discussion of how a foetus in the mother's womb can survive if the woman is struck by lightning and dies, the comparison is made with how a barrel of wine can be hit without the wine being spilled, because, apparently, the wine develops a protective skin (1458–1596). In further analogies, a pregnant woman is likened to a shoe around a foot, a sheath around a sword and a purse around pennies, the vessel in each case less valuable than the content, and explicitly described as weaker than what it protects. The analogy between a woman and a container is paralleled by one between maleness and the valuable content: the wine, foot, sword or money. In the questions following this section, the embryo is explicitly construed as a perpetuation of *man*'s seed that is delivered into the woman's body. No longer is any mention made of any female contribution to the embryo. Changes to the man's seed at conception *are* changes to the embryo. For instance, the question is raised in the text whether a seed that was intended to become a boy can be degraded and become a female embryo through the influence of a harmful external factor, as in the case of lightning striking the seed at the moment of conception (1569–73; 1593–6).

This implies both that the man's seed is perceived here to be the sole constituent of the foetus (while the woman's seed is not mentioned), and that the offspring is male by default. Various astrological predictions about the children's future, for example, the kind of beard they will grow, also automatically assume that they will be male. Women as vessels for these men are imagined to be leaking and to possess dangerous powers.

But it is again menstrual blood that poses the greatest danger to the foetus. Menstrual bleeding is associated with the death of an embryo, because when a pregnant woman suddenly menstruates, this shows that the pregnancy is over and the foetus has died (2140–4; 2649–51). If conceived during menstruation, children as a rule become leprous (2383).[9] Even after birth, babies and men are not safe from menstrual blood. Children in their cradle are poisoned by women's evil eye: 'It should be noted and kept in mind that the old women who thus have retained their period, and also other women whose menstrual fluid is fresh, poison the children when they are looking at their child lying in the cradle' (2315–18).[10] Here it is explained, in accordance with Aristotelian science, how the 'unclean matter' that is menstrual blood seeps out through the sweat holes in women's eyes, affecting the air around them, and then enters the children through the pores in their own eyes. Even the cradle is thus not protected against violation by menstrual blood. Grown men can be equally polluted: speaking to a menstruating woman, for instance, makes the distinctive male voice hoarse, again through the transmission of 'unclean' air (2377–82). But the greatest danger to men is contact with menstrual blood during sexual intercourse. That the penis during sex is the epitome of male vulnerability is illustrated by the warning to beware of women, especially prostitutes, who stick iron into the penis (584–94). This is particularly dangerous because the penis is supposedly the crossroads of all veins, and injuring it in this way means that menstrual blood can potentially enter the system (612–26). Sex with a menstruating woman also brings for men the risk of contracting leprosy and other incurable diseases (632–6). If stored in the body, the *Secreta* explains, menstrual blood causes women to crave strange food as well. These appetites are especially high in pregnancy, when the venomous menstrual blood is not regularly emitted, but locked into a woman's body: 'This is why a woman who is with child in many cases has food cravings, because during this time, the women's stomach is poisoned by venomous fluids, because the women's seed is enclosed in the mother's body near the stomach' (2155–8). These cravings can be for anything from apples to testicles, in the latter case preying directly on men's bodies (2632–47; 2160–6).

But what menstrual blood makes women hanker after in particular is men's more refined blood, semen. Driven by their menstrual fluids, women 'steal' semen by sucking it out with their vaginas during intercourse, and then retain it by letting themselves be impregnated (2093–7). Men are left deprived of their precious hot and moist fluid, which means that their health suffers and they die early:

> And thus the seed, being further ejected by the men, drains the body from which it is discharged, for the seed has the power to make moist and warm. But when the body dries up and the moisture is extracted from the body, there occurs an illness of body and life, and thus the human being dies. And this is why those [men] do not live long and die early who often and vigorously fornicate. (2909–14)

So here women do not just emit their own fluids and pollute men with them, but, driven by their voracious menstrual blood, they also extract fluids from men during sex.[11] But in each case it is menstruation which both defines and undermines the bodily relations between men and women. What the *Secreta* presents us with, then, are not curious superstitions, but accusations and fears that make sense within the world of this text. Even vile and far-fetched claims that women stain mirrors, kill children or suck out men's life blood are understandable in the context of the text's gender economy, in which men contain precious fluids. Ideas about menstruation are thus always specific to their historical context. Indeed, it would be wrong to presume that the Aristotelian position taken by the *Secreta* is representative of the attitudes of all medieval writers and people, just like there is no universal 'modern' view of menstruation.[12] But the *Secreta* does represent an influential strand of medieval thought in which gender is constructed bodily through the difference between menstruating and non-menstruating bodies.

Notes

1. For introductions to medieval medical ideas about menstruation see Bullough and Brundage; Cadden; Jacquart and Thomasset; and Kruse. For a more extensive discussion of the *Secreta*, see my *Medieval Blood*.
2. For the extent to which medical and natural philosophical theories were known outside the medical profession and the overlap between theology and natural philosophy see Bynum; Robertson; Wood; Ziegler, 'Practitioners and Saints' and ' "Ut dicunt medici" '.
3. The only other edited German translation was written c. 1460–65 by Johann von Hartlieb, physician at the court of Duke Albrecht III of Bavaria-Munich and his son Siegmund, and edited by Kristian Bosselmann-Cyran. The Latin

version is so far only accessible in modern publication in Helen Lemay's selective English translation of sixteenth-century prints. On the *Secreta* see also Schleissner, 'A Fifteenth-Century Physician's Attitude,' 'Pseudo-Albertus Magnus', 'Secreta mulierum'; and Sherwood-Smith. All English translations are my own.

4. Pliny, vol. 2, 548 (bk 7, chapter 64), Aristotle, 'On Dreams' 356–7. For more on medieval links between menstruation, mirrors and related motifs see Gaignebet.

5. See also Biller, 'A "Scientific" View' and 'Views of Jews'; and Johnson.

6. See Aristotle, *Generation of Animals* and Galen.

7. On gendered power and desire in this use of secrecy in the *Secreta* see Lochrie, 118–34.

8. McCracken has very astutely observed a similar distinction between male and female blood in medieval romance, by which only men's blood can carry positive significance. Men's blood has significance as sacrifice, as a sign of prowess or of pure lineage, while women's blood (that is, paradigmatically menstrual blood) is always suspect and excluded from positive public valuation in these romances (*Curse of Eve, passim*).

9. On medieval and early modern fears that a child conceived during menstruation will be sick or monstrous see Thomasset; and Niccoli. For examples of the survival of such ideas of women as leaking containers in general see Battersby; and Walker, 15–19.

10. On the evil eye see Salmon and Cabré.

11. For more on this sucking out of blood see my 'Bloodsuckers'. For more on general fears of menstrual pollution in medieval Christian thought see Elliott.

12. For more positive medieval views of menstruation, see Cadden, 70–9; McCracken, 'The Curse of Eve'; and 174; *Trotula*, 19–22. For a corrective to the assumption that menstruation is seen as polluting in all cultures, see Buckley and Gottlieb.

Works cited

Anon. 'Pseudo-Albertus Magnus, "Secreta Mulierum Cum Commento, Deutsch": Critical Text and Commentary'. Ed. Margaret Rose Schleissner. Diss. Princeton, 1987.

Aristotle. *Generation of Animals*. Ed. and trans. A. L. Peck. London: Heinemann, 1963.

——. 'On Dreams'. *On the Soul, Parva Naturalia, On Breath*. Ed. and trans. Walter Stanley Hett. Loeb Classical Library 288. London: Heinemann, 1957. 348–71.

Battersby, Christine. *The Phenomenal Woman: Feminist Metaphysics and the Patterns of Identity*. Oxford: Polity, 1998.

Bildhauer, Bettina. 'Bloodsuckers: The Construction of Female Sexuality in Medieval Science and Fiction'. *Consuming Narratives: Gender and Monstrous Appetites in the Middle Ages and the Renaissance*. Ed. Liz Herbert McAvoy and Teresa Walters. Cardiff: Wales University Press, 2002. 104–15.

——. *Medieval Blood*. Cardiff: Wales University Press, 2005.

Biller, Peter. 'A "Scientific" View of Jews from Paris around 1300'. *Micrologus* 9 (2001): 137–68.

——. 'Views of Jews from Paris around 1300: Christian or "Scientific"?' *Christianity and Judaism: Papers Read at the 1991 Summer Meeting and the 1992 Winter Meeting of the Ecclesiastical History Society*. Ed. Diana Wood. Oxford: Blackwell, 1998. 187–207.

Bosselmann-Cyran, Kristian, ed. *'Secreta mulierum' mit Glosse in der deutschen Bearbeitung von Johann Hartlieb [Secreta mulierum with commentary in the German version by Johann Hartlieb]*. Pattensen/Hannover: Wellm, 1985.

Buckley, Thomas and Alma Gottlieb. 'A Critical Appraisal of Theories of Menstrual Symbolism'. *Blood Magic: The Anthropology of Menstruation*. Ed. Thomas Buckley and Alma Gottlieb. Berkeley: California University Press, 1988. 3–50.

Bullough, Vern L. and James A. Brundage, eds. *Handbook of Medieval Sexuality*. New York: Garland, 1996.

Bynum, Caroline Walker. 'The Female Body and Religious Practice in the Later Middle Ages'. *Fragmentation and Redemption: Essays on Gender and the Human Body in Medieval Religion*. Ed. Caroline Walker Bynum. New York: Zone, 1991. 181–238.

Cadden, Joan. *Meanings of Sex Difference in the Middle Ages: Medicine, Science, and Culture*. Cambridge: Cambridge University Press, 1993.

Douglas, Mary. *Purity and Danger: An Analysis of the Concepts of Pollution and Taboo*. London: Routledge, 1966.

Elliott, Dyan. *Fallen Bodies: Pollution, Sexuality and Demonology in the Middle Ages*. Philadelphia: Pennsylvania University Press, 1999.

Fisher, Will. 'The Renaissance Board: Masculinity in Early Modern England'. *Renaissance Quarterly* 54 (2001): 155–87.

Gaignebet, Claude. 'Véronique ou l'image vraie.' *Anagrom* 7–8 (1976): 45–70.

Galen. *On Seed*. Vol. 4 of *Opera Omnia*. Ed. Carl Gottlob Kühn. 20 vols. Leipzig: Cnobloch, 1821–33.

Green, Monica H. 'From "Diseases of Women" to "Secrets of Women": The Transformation of Gynecological Literature in the Late Middle Ages'. *Journal of Medieval and Early Modern Studies* 30 (2000): 5–39.

Jacquart, Danielle and Claude Thomasset. *Sexuality and Medicine in the Middle Ages*. Trans. Matthew Adamson. Princeton: Princeton University Press, 1988.

Johnson, Willis. 'The Myth of Jewish Male Menses'. *Journal of Medieval History* 24 (1998): 273–95.

Kruse, Britta-Juliane. *Verborgene Heilkünste: Geschichte der Frauenmedizin im Spätmittelalter [The Hidden Art of Healing: History of Women's Medicine in the Late Middle Ages]*. Berlin: de Gruyter, 1996.

Lemay, Helen Rodnite. *Women's Secrets: A Translation of Pseudo-Albertus Magnus's 'De Secretis Mulierum with Commentaries'*. Albany: State University of New York Press, 1992.

Lochrie, Karma. *Covert Operations: The Medieval Uses of Secrecy*. The Middle Ages Series. Philadelphia: Pennsylvania University Press, 1999.

McCracken, Peggy. 'The Curse of Eve: Female Bodies and Christian Bodies in Heloise's Third Letter'. *Listening to Heloise: The Voice of a Twelfth-Century Woman*. Ed. Bonnie Wheeler. New York: St Martin's, 2000. 217–31.

——. *The Curse of Eve, the Wound of the Hero: Blood, Gender, and Medieval Literature*. Philadelphia: Pennsylvania University Press, 2003.

Niccoli, Ottavia. ' "Menstruum Quasi Monstruum": Menstrual Births and Menstrual Taboo in the Sixteenth Century'. *Sex and Gender in Historical Perspective*. Ed. Edward Muir and Guido Ruggiero. Baltimore: Johns Hopkins University Press, 1990. 1–25.

Pliny (Plinius Cacilius Secundus). *Natural History*. Ed. Harris Rackham, trans. Harris Rackham *et al*. 10 vols. Loeb Classical Library. London: Heinemann, 1938–62.

Robertson, Elizabeth. 'Medieval Medical Views of Woman and Female Spirituality in the *Ancrene Wisse* and Julian of Norwich's *Showings*'. *Feminist Approaches to the Body in Medieval Literature*. Ed. Linda Lomperis and Sarah Stanbury. Philadelphia: Pennsylvania University Press, 1993. 142–67.

Salmón, Fernando and Montserrat Cabré. 'Fascinating Women: The Evil Eye in Medical Scholasticism'. *Medicine from the Black Death to the French Disease*. Ed. Roger French *et al*. Aldershot: Ashgate, 1998. 53–84.

Schleissner, Margaret Rose. 'A Fifteenth-Century Physician's Attitude Toward Sexuality: Dr Johann Hartlieb's *Secreta mulierum* Translation.' *Sex in the Middle Ages: A Book of Essays*. Ed. Joyce Salisbury. New York: Garland, 1991. 110–25.

——. 'Pseudo-Albertus Magnus, *Secreta mulierum*: Ein spätmittelalterlicher Prosatraktat über Entwicklungs- und Geburtslehre und die Natur der Frauen'. ['Pseudo-Albertus Magnus, *Secreta mulierum*: A late medieval prose treatise on embryology and obstetrics and the nature of women'.] *Würzburger medizinhistorische Forschungen* 9 (1991): 115–24.

——. 'Secreta mulierum'. *Verfasserlexikon*. Vol. 8 of *Die deutsche Literatur des Mittelalters*. Ed. Kurt Ruh. 11 vols. 2nd ed. Berlin: de Gruyter, 1978. Col. 986–93.

Sherwood-Smith, Maria. 'Forschung oder Vorurteil, Kultur oder Naturkunde?: Zur Frage der Frauenfeindlichkeit in den deutschen und niederländischen Bearbeitungen der Secreta Mulierum von Pseudo-Albertus Magnus.' ['Research or Prejudice, Culture or Natural Science? On the Issue of Misogyny in the German and Dutch Versions of the Secreta Mulierum by Pseudo-Albertus Magnus'.] *Natur und Kultur in der deutschen Literatur des Mittelalters: Colloquium Exeter 1997*. Ed. Alan Robertshaw and Gerhard Wolf. Tübingen: Niemeyer, 1999.

The Trotula: An English Translation of the Medieval Compendium of Women's Medicine. Ed. and trans. Monica Green. Philadelphia: Pennsylvania University Press, 2001.

Thomasset, Claude. 'La Femme au Moyen-Âge: Les composantes fondamentales de sa répresentation: immunité – impunité'. ['Woman in the Middle Ages: The Fundamental Components of Her Representation: Immunity–Impunity'.] *Ornicar* 22–3 (1981): 223–38.

Walker, Anne. *The Menstrual Cycle*. London: Routledge, 1997.

Wood, Charles T. 'The Doctors' Dilemma: Sin, Salvation and the Menstrual Cycle in Medieval Thought'. *Speculum* 56 (1981): 710–27.

Ziegler, Joseph. 'Practitioners and Saints: Medical Men in Canonization Processes in the Thirteenth to Fifteenth Centuries'. *Social History of Medicine* 12 (1999): 191–225.

——. ' "Ut dicunt medici": Medical Knowledge and Theological Debates in the Second Half of the Thirteenth Century'. *Bulletin of the History of Medicine* 73 (1999): 208–37.

6
Menstrual Knowledge and Medical Practice in Early Modern France, c. 1555–1761

Cathy McClive

'What is menstrual blood?' and 'what is the function of menstrual blood?', asked Marguerite du Tertre de la Marche, *maîtresse sage-femme en chef* (chief mistress-midwife) of the Hôtel-Dieu, Paris, of her apprentice midwives and other readers in her manual of 1677 (19). Following the question-and-answer format frequently used in classical and medieval works and more recently revitalised in the form of the catechism by the Council of Trent (1545–63), du Tertre provided her own definition:

> Menstrual blood is that which is expelled by the matrix every month in healthy women who are neither pregnant nor breast-feeding ... it is natural for the woman to shed blood every month, between the ages of fourteen or fifteen and forty-five or fifty, more or less. It is also natural that she shed menstrual blood after childbirth: on a monthly basis the blood is called monthlies, ordinaries, flowers or menses, and after childbirth, evacuations. But when the evacuations are excessive, that is called loss of blood, as when the woman is pregnant. (19, 45)[1]

The inclusion of this particular question in a midwifery text reveals the importance accorded to understanding the function of menstrual bleeding in order to identify the cause of the expulsion of bloody fluids by the female body at various points during the reproductive process.[2] At first glance du Tertre's definition of menses appears recognisable to the modern reader – menstrual blood is seen to flow monthly in non-pregnant, non-lactating women between the ages of roughly fourteen and fifty. Things become more complicated, however, with the inclusion of post-partum bleeding and bleeding during pregnancy, or spotting, in the generic definition of menstrual blood. du Tertre interpreted a variety of bloody emissions as menstrual blood. The terms she used – 'monthlies, ordinaries, flowers and menses' for naturally-occurring monthly emissions, 'evacuations' for post-partum bleeding, and 'blood-loss' for copious lochia (heavy post-partum bleeding) and bleeding during pregnancy – reveal a set of expectations about when and in what conditions the female body was expected to bleed and, more significantly, that the blood expelled in all these emissions was considered to be menstrual. It thus seems that what 'mattered' most in terms of menstrual bleeding in early modern French medicine was the substance produced by a body, over and above what we would understand as the physiological *process* implicated in the expulsion of the fluid, involving ovulation, the thickening of the endometrium and its eventual shedding.

Early modern perceptions of menstrual blood and bleeding thus shaped, and were shaped by, the linguistic and cultural merging of various bloody, female flows into one substance. The plethora of terms available to describe menstrual blood emitted naturally every month outside pregnancy and lactation suggests that perceptions of the type of flow were heavily dependant on the circumstances and condition of the body when it flowed. The substance itself was seen as the same, although its emission had different consequences depending upon the context. The merging of what we today would recognise as different physiological processes and substances also impacted on understandings of the link between menses and reproduction. Although it was a commonplace understanding that menstrual bleeding ceased during pregnancy, exceptions to this rule were observed and referred to as emissions of menstrual blood, as du Tertre noted. The plasticity of such early modern interpretations of menstrual bleeding is accentuated by the fact that it was not possible for du Tertre to ask her apprentices to define 'menstruation' as a process – rather than 'menstrual blood' as a substance – since the term 'menstruation' is not documented as having been used in French until a century later. The first occurrence of the term 'menstruation' in a French-language medical text dates from 1761 when it was most probably introduced by the physician Jean Astruc in his *Traité des maladies des femmes*. Astruc's treatise was culled from a series of medical lectures on women's diseases that he had given at least two decades earlier, suggesting that the term 'menstruation' may have been in use before being set into print.[3] However, the transition was not uniform since even Astruc's text refers interchangeably to 'menstruation' (*menstruation*), 'rules' (*les règles*), 'menstrual flux' (*le flux menstrual*) and 'periods' (*les retours périodiques*). Moreover, the word 'menstruation' did not appear in French-language dictionaries until 1932, following the discovery of ovulation and hormones, despite the occurrence of *menstrues and menstrual*, derived from the same Latin and Greek etymology, much earlier (McClive, 'Bleeding Flowers' 21).

An historical study of early modern menstrual vocabulary has important implications for our understandings of contemporary perceptions of the sexed body and womanhood. Danielle Jacquart, Claude Thomasset and later Thomas Laqueur have argued that the lack of specific nomenclature in Latin for female genitalia in the pre-modern period contributed considerably to the development of the interchangeable 'one-sex model' (Jacquart and Thomasset 15–25, 35; Laqueur, *Making Sex* 96). Laqueur follows Aristotelian teleology and Galenic humouralism into depictions of man and woman as anatomically and physiologically homologous. Woman was seen as an imperfect version of man with an inverted, internal set of (male) genitalia, as she lacked sufficient heat to push them outwards. Male and female secondary sexual characteristics were a matter of degree on a sliding humoural scale of hot and cold. As such, Laqueur argues, this 'natural' model of sexual homology became a culturally-embedded reference with the emphasis placed on gendered or 'cultural', rather than biological sexual difference (Laqueur, *Making Sex*).[4] A dramatic model of change, around 1750 to 1760, cited frequently by advocates of the 'one-sex model', has to a large extent dominated the historiography of menstrual bleeding (Laqueur, 'Sex in the Flesh' 306). Historians have argued that there was a significant shift in medical perceptions of menstrual bleeding at this juncture and that henceforth the definition of menstrual bleeding was limited to the expulsion of blood from the vagina alone, ruling out the possibility of the inclusion of spotting, vicarious and male periodic bleeding in the category of menses (Lord 41; Stolberg, this collection).

Within this post-1750 paradigm, the newly defined process of menstruation (in French at least) belonged to women only and symbolised their fertile potential. Menstruation tied women more closely to their reproductive nature, introducing the notion of menstrual bleeding as an essential signifier of the female sex, and as a precise process, rather than the more flexible view of menses as one humoural flux among a mass of homogenous bodily fluids.[5] This would seem to fit with the codification of the term 'menstruation' in French medical vocabulary in 1761. This model of paradigm shifts in the perception of menstrual bleeding is dangerously seductive, offering a simple yet reductionist explanation for the divergence between modern and pre-modern socio-cultural and medical interpretations of menstrual bleeding and sexual difference. It has been widely discussed in terms of the nature/culture (essentialism/ constructivism) debate in women's and gender history whereby feminist scholars have decried the perpetuation of the 'natural' equation of woman with her reproductive function and of man with 'culture' (Butler 66; Fisher 155–6, 184–5; Jordanova 15). For instance, scholars such as Patricia Crawford, Alexandra Lord and Gail Kern Paster have harnessed the 'natural' medical explanations for the demarcation of sexual difference in seeking to account for the socio-economic inferiority of early modern women (c. 1450–1750) and for 'scientific' preoccupations with the pathologising of menstrual bleeding. Precluding analysis of the full complexity of early modern perceptions of menstrual bleeding, this model has also been criticised by classical and medieval historians who demonstrate the insufficiency of binaries to account for the pre-modern enigma of menses, and who argue that the assimilation of male discharges with menstrual bleeding does not mean that menses were always perceived as being gender-neutral prior to 1750 (Cadden 3; King 7–11, 15; Stolberg, this collection).

In order to unpick the tangle of cultural assumptions which have led to the pre- and post-1750 models of menstrual bleeding in some historical accounts of early modern medicine it is helpful to look at menses in the wider picture of humouralism and early modern bodies in general. This effectively emancipates an analysis of attitudes to menses from the issue of gender, and facilitates the exploration of attitudes to menstrual blood in terms of other kinds of bloody emission. It also refocuses the discussion on the material substance of menstrual blood in line with early modern French perceptions of menses, as well as underlining the practical impact of the construction of menstrual blood as an all-encompassing term and matter. The examples of vicarious menstrual bleeding and the dual physiology of menstrual bleeding during pregnancy illustrate particularly well the ambiguity of early modern perceptions of menstrual bleeding and the significant role played in such uncertainty by the linguistic emphasis on menstrual matter over and above process.

Menses and blood were inherently linked in the hierarchical humoural economy of fluids (blood, phlegm, yellow bile and black bile). Early modern medical practitioners were in agreement about the importance of blood – the most vital of all the humours and the source of life itself. Blood was not sexed within the humoural economy of fluids, but was held to ebb and flow in both male and female bodies. Excess humour caused ill-health in male and female bodies and a body naturally tended towards the evacuation of such superfluous matter. The production of blood, and indeed other bodily fluids such as menses, semen, and milk, occurred during various stages of 'cooking'. The analysis of blood and bloody fluids provided an excellent means to diagnose health and the humoural equilibrium of the body (Long 225–68). Hence, the material used for the manufacture of surgical basins was carefully chosen to preserve the collected blood in optimum conditions for diagnosis (Long 251). Once coagulated, blood readily revealed which humour was lacking or was in excess according to the colour and texture of the dried matter. Similarly, linen soaked in menstrual blood and left to dry also offered insight into the health of a female patient (Courtin 259). Menstrual matter, like other bloody fluids, was not perceived as impure *per se* in early modern France. Rather, diagnoses of the condition of expelled menses depended on the health and condition of the menstruating individual. Many elite, learned physicians such as Jean Liébault (1535–96) argued that the vices and malefic properties attributed to menses by the ancients resulted from 'the mixing of some vicious and corrupt humour, or because of the bad condition of the uterus' rather than the condition of the menses, of which he declared 'there is nothing so benign in the human body as blood which is the treasure of life itself' (536–7). In certain circumstances menses were attributed with salutary, therapeutic powers alongside their natural purgative properties. Jurist André Tiraqueau (fl. 1535), for instance, argued that menses could cure epilepsy, paralysis and ease the pains of childbirth (136–7).

Not only were blood and semen and blood and milk linked in humoural theory, various types of bleeding or bloody discharge were also indistinguishable in both real and cultural terms. As we have seen, many emissions which we now regard as entirely separate, such as postpartum bleeding and spotting, were defined as menstrual blood. If, however, the regular route used for evacuation was obstructed, it was believed that the fluid risked remaining in the body. In such cases the body would attempt to evacuate the retained menstrual fluid via a different orifice. This could include the nose, the anus, the arteries, the mouth and the pores. Significantly, these vicarious emissions were assimilated linguistically and culturally with menstrual bleeding and, although generally held to be a female condition, were also applied to male bodies. In cases of failed menstrual bleeding it was equally acceptable for the body to expel the matter via an alternative orifice as it was for this process to be artificially induced by a physician or surgeon in the form of blood-letting or purging. Royal physician Jacques Dubois (1478–1555) noted: 'When the monthlies fail, if the blood flows from the nostrils it is a good thing, but more so if it flows from another part, either through nature's work or through the operation of a physician, it is healthy' (160). In certain contexts, therefore, blood-letting was culturally assimilated to menstrual bleeding in the same way that a nose-bleed could represent a deviated spontaneous alternative. Male menstrual bleeding or vicarious evacuation was generally perceived to occur through the anus or genitals, unless it was induced in the form of phlebotomy (blood-letting). Haemorrhoidal bleeding could provide relief and balance for the male humoural body unable to evacuate excess liquid through more usual means (McClive, 'Bleeding Flowers' 256–64; Pomata 124–35). Within the humoural system, warmer, drier men more easily transformed excess fluid into bodily hair or perspiration than their cooler, wetter, female counterparts. Saturnine or splenetic men who did not have regular haemorrhoidal fluxes were condemned to suffer blockages in the spermatic vessels and in the veins in their thighs (de Blégny, *Fragment* 35). Nuns and widows were also thought to be subject to haemorrhoidal flux because the lack of conjugal activity caused obstructions to develop in the uterine and vaginal vessels forcing superfluities to exit elsewhere (Guyon 250).

Despite the obvious health-giving benefits, vicarious menstrual bleeding posed a serious problem for medical practitioners. Arguably, the most important thing to consider was that the body should be relieved of its excess blood or other humour. In certain cases it was preferable that the practitioner intervene and attempt to restore the natural course of the flow. Provincial physician Louys Guyon (d. 1630) argued that such flows indicated an over-abundance of blood in the body, particularly when they stemmed from the ears or nostrils. He advocated blood-letting to encourage the flow to return to its natural pathway and specific treatments depending on the location of the flow. If the right nostril was the source of flow for instance, then the patient should be bled from the arm and clysters (plasters or compresses used to draw out bad humours and unblock obstructions) applied to the liver (121–5). On the other hand, if the patient was plethoric (naturally disposed towards an excess of humours), such measures could be dangerous, arresting the flow prematurely and causing it to stagnate in the body. It was the practitioner's responsibility to examine the quality and quantity of the expelled blood and to determine when to intervene and when to step back and let nature run its course. If the patient was plethoric and the liquid was clear and pure then bleeding could be beneficial. If the flow was impetuous, rather than regular, this was a bad sign suggesting illness rather than a natural periodic evacuation of excess fluid. If the quantity expelled exceeded four *livres* (roughly two litres) this too was dangerous and the practitioner should intervene to arrest the flow (Guyon 207). When the menses were suppressed,

> in such cases we should not try the method of revulsion through the parts indicated for this by nature ... but to those who are subject to such natural purging, we should let the flux continue within reason, as with haemorrhoids, vomiting of blood, the opening of veins at certain quarters of the moon or certain seasons as I've seen often. (Guyon 126)

For Guyon, there should be no interference with other flows when the menses stopped.

In February 1680 surgeon Nicolas de Blégny (1652–1722) published a letter from one Belin, a provincial physician from Vasian, to royal physician abbot Bourdelot, in one of his medical journals. This letter concerned a girl who menstruated from her eyes. At eighteen years of age the girl was short, thin and experienced 'a loss of blood from her eyes which flowed nearly continuously in the manner of tears causing pain and inflammation' on a monthly basis. The bleeding lasted for four days at a time and returned the following month. Belin observed the phenomenon over two months. He wrote that, noting 'a periodic and regulated recurrence of this indisposition, I did not doubt that it was menstrual, having found an obstacle in the matrix, the menses had flowed to the brain and from there were expelled' (de Blégny, *Temple* 69). On learning that the girl had never menstruated in any other manner, Belin prescribed bleeding from the arm and foot followed by an herbal remedy. He proclaimed his successful diagnosis and intervention, recording that 'her ordinaries were happily induced and she has had them regularly ever since at the same hour and time of the month' (de Blégny, *Temple* 69). In this case the physician was alarmed by the fact that the girl had only ever menstruated from her eyes and feared that if uncorrected the body would retain this habit into adulthood, prolonging the painful monthly symptoms she endured and seriously affecting her future fertility. In this sense, although the menstrual flow fitted perfectly the regular, orderly pattern of colour, quantity and timing, her monthly bleeding appeared to be permanently, rather than unusually, diverted. Hence, there was a need to re-train the body's flows to take the path of genital menstrual bleeding so that she might later conceive and bear children. Vicarious menstrual bleeding served the same purpose as 'menstruation' and indeed effected the expulsion of superfluous menstrual matter from the body; however, it precluded the potential to procreate for which the presence of menses in the uterus was necessary.

Many medical practitioners observed the phenomenon of menstrual bleeding during pregnancy and sought to reconcile this occurrence with the commonplace notion of the role of menses as nourishment for the foetus during gestation and the cessation of menstrual flux as, an albeit ambiguous, sign of conception (McClive, 'Hidden Truths' 209–13). Early modern medical practitioners conceived of female and male physiology in an entirely different manner to the way in which we understand it today. Whereas we understand a series of different and entirely separate physiological processes, producing different bodily fluids, early modern medicine recognised one process and one fluid, although this could take varying routes. For instance, as we have seen, various different types of bleeding were grouped together under the generic term 'menstrual'. By contrast, there also existed a series of separate physiological models or channels within the humoural economy to account for the flow of blood in the bodies of virgins, widows and pregnant women. The female body was thus perceived to have two potentially different physiological mechanisms which could be utilised according to the age and condition of the individual woman, although both were included in the linguistic and cultural category of menstrual bleeding as they performed the same function. This notion stemmed from the belief that menses were stored in the uterus to nourish the foetus during pregnancy and that the body would seek to expel any superfluity without endangering the life of the foetus. Thus, during pregnancy, menstrual bleeding was generally perceived to derive from the vagina rather than the uterus. The linguistic ambiguity surrounding menstrual bleeding is complicated for the modern reader by the sometimes confusing and interchangeable anatomical terminology of the early modern medical world (Jacquart and Thomasset 10, 25–7). Scientific, and particularly anatomical, vocabulary was still somewhat flexible and different terms could be used to mean the same thing, or a single term could be used with regard to several parts of anatomy. Such ambiguity exists, for example, in the expressions 'mouth of the uterus' and 'neck of the uterus', which could be interpreted to signify either the cervix or the vagina. Early modern medical practitioners also wrote of 'orifices' in the depths of the uterus as well as at its entrance. It is not always clear, therefore, whether a practitioner was referring to the vaginal entrance, cervix or womb, in relation to the provenance of menstrual flow.

Medical practitioners concentrated on analyses of the quality and quantity of menstrual blood in order to determine whether it had been evacuated through the uterine or vaginal vessels. Two different types of menses were described depending on the provenance of the flow.

Dubois observed that a darker fluid flowed from the depths of the womb whilst a lighter fluid originated from the vessels nearer to the vaginal entrance (136). He also argued that virgins and post-partum women menstruated from the vessels in their thighs and legs which travelled to the matrix (158). Liébault argued that there were two types of menstrual fluid, stemming this time from the hypogastric and the spermatic veins. Fluid from the hypogastric veins originated in the vagina and was expelled by virgins and pregnant women whereas fluid from the spermatic vessels came from the depths of the womb and was evacuated by fertile, non-pregnant women (333). Du Tertre made a similar distinction between the path taken by the menstrual flow of non-pregnant girls and women, and the preferred route of the menstrual fluid in expecting mothers. She argued that menstrual blood was contained in the veins and arteries of the uterus, as well as in the uterine cavity itself. These uterine vessels stemmed from the hypogastric and spermatic canals which fed into both the 'neck of the uterus' and the uterine cavity: 'from the vessels in the depths of the matrix the blood flows in girls and women who are not pregnant, and it happens that sometimes pregnant women have their monthlies, and in this case the blood flows from the vessels at the neck of the matrix' (19–20). Miscarriage was thought to occur if the preferred route in such circumstances (via the vaginal vessels) was obstructed in a pregnant woman and the blood had no option but to flow through the uterus carrying the foetus with it.

There were two methods of judging the provenance of the fluid. Midwives and *accoucheurs* (male midwives) tended to prefer internal cervical examinations and an analysis of the colour, texture and quality of the emission. For instance, du Tertre tested whether the menstrual fluid stemmed from the 'neck or depths of the matrix' by performing a careful internal examination of the vagina and cervix and an analysis of the quality of the fluid:

> I will judge it by the colour and consistency of the blood and also the manner in which it flows. When the blood flows from the depths of the matrix, it is black and coagulated because of the length of time it has spent there and flows abundantly and the internal orifice is open; but when it exits from the neck of the matrix, it is crimson and fluid because it does not stay long in this part and does not flow abundantly and the internal orifice is entirely closed. (47–8)

Physicians, surgeons and anatomists on the other hand tended to rely more heavily on evidence from dissections of women who had died during menstrual bleeding or pregnancy to illustrate the path of menstrual bleeding. Physician Jacques Duval (1555–1620) quoted the findings of his contemporary Pinean, a Parisian master surgeon. Pinean dissected cadavers of women who had died during their menstrual flux, whence he observed that the vessels in the vagina were 'full and swollen with blood, the rest of the matrix pure and clean from this moisture and bloody excretion' (90). Duval concluded from this observation that menstrual blood 'does not flow from the uterus but from the anasto-mose or dilation of the orifice of these vessels, which is in fact the channel of the vulva' (90). He used this evidence to support the dual physiology of pregnant women and remarked that one should not be alarmed when a pregnant woman menstruated, as the vaginal flow was unlikely to harm the foetus, safely ensconced in the uterus.

In 1702 M. Littré exposed his own research in a letter to the *Académie Royale des Sciences*. In line with the growing emphasis on empiricism and anatomical proof, Littré accentuated his own first-hand experience and observations of the route taken by menstrual blood:

> I think that we can say that the blood of periods flows from the walls of the matrix and not from the vagina. Such observations as I have made from several girls and women who died during the time of their periods confirm this judgement. But the three observations that I have made on a girl and two women remove any doubt. All three had a descent of the matrix, in each one the internal orifice appeared at the level of the large lips of the vulva. I remarked in all three that the menstrual blood flowed out of the internal orifice of the matrix and that there was not a single drop in the vaginal cavity itself. (Planque 139–40)

Such direct observations were used by physicians like Jean Palfyn (1650–1730) to contend that menstrual bleeding only stemmed from the vagina when all other routes were obstructed, or if the woman in question was pregnant. He reasoned that if menstrual bleeding derived from the depths of the uterus during pregnancy, women would con-stantly miscarry. Instead, during gestation, menstrual bleeding could pass through the part of the uterus not touching the placenta, or through the vaginal canals, to ensure that the woman and her foetus were not inconvenienced (113). It is vital to situate any analysis of men-strual bleeding within the wider context of contemporary constructions of the body and health. In early modern France the prevailing view of the body remained humoural; it is therefore logical that menstrual bleeding should have been primarily conceived of within this framework. Within this context of humouralism, menstrual blood and menstrual bleeding, both substance and process, offered generic terms for a multitude of bloody fluids and discharges. More significantly, such 'menstrual' emissions were defined as a result of the substance evacuated over and above the orifice from which the fluid was expelled and the process leading to the emission. This meant that vicarious bleeding and specifically male periodic bleeding could be assimilated to 'menstruation'. Moreover, as with bleeding during pregnancy, analyses of the menstrual matter were employed to account for the origins of the flow and thus the process which had produced it.

Caution is necessary however. Early modern constructions of the sexed body and womanhood were complicated by the notion that menstrual bleeding could occur in a multitude of ways and was not necessarily restricted to non-pregnant, non-lactating women, but could also be experienced by men, by expectant mothers and through various alternative routes. Thus, menstrual bleeding was not always perceived as an essential, natural signifier of the sexed body and womanhood in early modern French medicine – yet neither was it entirely ungendered. The link between menses and reproduction, although tenuous, existed and is revealed in the examples of vicarious menstrual bleeding and the dual physiology of pregnancy outlined in this article. Medical practitioners' attempts to retrain permanently-diverted vicarious menses in mature pre-menopausal women to ensure their fertile potential underline the association of menses and conception. The dual physiology attributed to pregnant women to account for foetal nourishment through retained menses in the uterus and simultaneous spotting emphasised not only the flexibility of female physiology but, more importantly, the understanding that menstrual fluid played a vital role in reproduction and that copious 'menstrual' bleeding during gestation could harm the unborn child.

The decade 1750–60 is frequently cited as pivotal in the history of menstrual bleeding. A refining of medical vocabulary and the tightening of anatomical and physiological definitions are claimed to have altered perceptions of womanhood and menstruation around this time and medical constructions of menstruation began to resemble those familiar to the modern reader. However convenient, this model is of limited historical value, eliding the full complexities and ambiguities of early modern socio-cultural and medical perceptions of the enigma of menstrual bleeding. Since it was not until the discovery of ovulation in the 1870s and hormones in the 1930s that the biological function we recognise as menstruation was fully revealed and integrated into French culture, it seems anachronistic to suggest that any decisive shift occurred prior to this. It is equally problematic to suggest that prior to 1750 any periodic bloody flow was interpreted as menstrual, and that post-1761 the definition of 'menstruation' was limited to vaginal bleeding in mature, pre-menopausal, non-pregnant women. Early modern perceptions of menstrual bleeding were ambiguous and enigmatic, covering what we would understand as a variety of fluids and physiological processes, and as such are not always easy for the modern historian of medicine to grasp. In order to better appreciate the polymorphic nature of menstrual bleeding in early modern France the historian must initially investigate perceptions of the material substance of menstrual blood, rather than seeking to understand the *process* of menstruation.

Notes

1. All translations are my own.
2. Preoccupation with menstruation in the context of midwifery occurred considerably earlier in France than in England (Lord 39).
3. We know that Astruc was lecturing on this topic in the early 1740s because one of Astruc's former students, an English physician, published his notes taken during these lectures in 1843.
4. For a contrasting view see Pomata, who subverts the male-dominated hierarchy of the 'one-sex' (131–47).
5. Pomata argues that menstruation was more firmly linked to generation by the discovery of ovulation in the late nineteenth century and of hormones in the 1930s (146).

Works cited

Astruc, Jean. *Traité des maladies des femmes, où l'on a tâché de joindre à une théorie solide la pratique la plus sûre et la mieux éprouvée avec un catalogue chronologique des médecins qui ont écrit sur ces maladies [Treatise of women's diseases where we have tried to join a solid theory to the surest and most proven experience, with a chronological catalogue of physicians who have written about these illnesses].* 6 vols. Paris: Guillaume Cavelier, 1761.

Butler, Judith. *Bodies that Matter: On the Discursive Limits of Sex.* New York: Routledge, 1993.

Blégny, Nicolas de. *Fragment d'un projet d'histoire concernant la chevalerie chrestienne, au sujet des remèdes exquis et des panacées charitables, envoyez à Cayenne l'an 1697 par les Hospitaliers du Saint-Esprit [Fragment of an historical project concerning Christian chivalry, on the subject of the exquisite remedies and charitable cure-alls sent to Cayenne in 1697 by the Hospitaliers of the Holy Spirit].* Angers: Veuve D. Avril, 1697.

——. *Le Temple Desculpe, ou le depositaire des nouvelles descouvertes, qui se font journellement dans toutes les parties de la medecine [The temple of Desculpe or the depository for the new discoveries happening daily in all sections of medicine]*. Paris: Claude Blageart and Laurent d'Houry, 1680.

Cadden, Joan. *Meanings of Sex Difference in the Middle Ages: Medicine, Science and Culture*. Cambridge: Cambridge University Press, 1993.

Courtin, Germain. *Leçons anatomiques et chirurgicales de feu me Germain Courtin, docteur regent en la faculté de medecine à Paris, receuillies par Estienne Binet, chirurgien juré à Paris [Anatomical and surgical lessons of master Germain Gourtin, regent phyisician in the Parisian faculty of medicine, collected by Estienne Binet, sworn surgeon of Paris]*. Paris: Denys Langlois, 1612.

Crawford, Patricia. 'Attitudes to Menstruation in Seventeenth-Century England'. *Past and Present* 91 (1981): 47–73.

Dubois, Jacques. *De l'utilité des moys des femmes [Of the usefulness of women's monthlies]*. Paris: G. Morel, 1559.

Duval, Jacques. *Les hermaphrodits. Accouchemens des Femmes, et traitement qui est requis pour les relever en santé et bien élever leurs enfans [Hermaphrodites. Childbed and the treatment required to bring them back to health and to raise their children]*. Rouen: David Geoffroy, 1612.

Guyon, Louys. *Les cours de medecine en françois, contenant le miroir de beauté et de santé corporelle par M. Louys Guyon, Dolois, Sr de la Nauche, docteur en médecine, et la théorie avec un accomplissement de la pratique par M. Lazare Meysonnier [Medical lectures in French, containing the mirror of beauty and physical health by Monsieur Louys Guyon, of Doles, Sieur de la Nauche, doctor of medicine, and the theory with a practical accomplishment by Monsieur Lazare Meysonnier]*. Ed. Lazare Meysonnier. 2 vols. 1673. Paris: Louis Pariente, 1987.

Jacquart, Danielle and Claude Thomasset. *Sexualité et savoir médical au moyen âge [Sexuality and medical knowledge in the middle ages]*. Paris: Presses Universitaires Françaises, 1985.

Jordanova, Ludmilla. *Sexual Visions: Images of Gender in Science and Medicine between the Eighteenth and Twentieth Centuries*. Madison: Wisconsin University Press, 1989.

King, Helen. *Hippocrates' Woman: Reading the Female Body in Ancient Greece*. London: Routledge, 1998.

Laqueur, Thomas. *Making Sex: Body and Gender from the Greeks to Freud*. Cambridge: Harvard University Press, 1990.

——. 'Sex in the Flesh'. *Isis* 94 (2003): 300–6.

Liébault, Jean. *Trois livres des maladies et infirmitez des femmes pris du Latin de M. Iean Liebaut docteur medecin à Paris [Three books on the illnesses and infirmities of women taken from the Latin of Monsieur Jean Liébalt, doctor of medicine in Paris]*. Rouen: Iean Berthelin, 1649.

Long. 'Traité des principes de chirurgie de Maître Long tirez par Iean Sue, étudiant en chirurgie à Paris le 3 janvier 1723' ['Treatise of the principles of surgery by Master Long taken down by Iean Sue, student in surgery at Paris 3 January 1723']. Ms. 5089. Bibliothèque Inter-Universitaire de la Médecine, Paris.

Lord, Alexandra. ' "The Great Arcana of the Deity": Menstruation and Menstrual Disorders in Eighteenth-Century British Thought'. *Bulletin of the History of Medicine* 73 (1999): 38–63.

McClive, Cathy. 'Bleeding Flowers and Waning Moons: A History of Menstruation in France c. 1495–1761'. PhD Diss. University of Warwick, 2004.

——. 'The Hidden Truths of the Belly: The Uncertainties of Pregnancy in Early Modern Europe'. *Social History of Medicine* 15.2 (2002): 209–27.

Marche, Marguerite du Tertre de la. *Instruction familiere et utile au sages-femmes pour bien pratiquer les accouchemens, faite par demandes et réponses [Familiar and useful guide for midwives in how to practise deliveries, in question and answer format]*. 1677. Paris: Laurent d'Houry, 1710.

Palfyn, Jean. *Description anatomique des parties de la Femme qui servent à la Generation avec un traité des monstres, de leur causes, de leur Nature, et de leur differences et une description anatomique de la disposition surprenante de quelques parties externes, et internes de deux enfants nés dans la ville de Gand, capitale de Flanders, le 23 avril 1703 [Anatomical description of the parts of a woman which serve generation with a treatise on monsters and their causes, their nature, and their differences and an anatomical description of the surprising disposition of several external and internal parts of two infants born in the town of Ghent, capital of Flanders, 23 April 1703]*. Leiden: Bastiaan Schouten, 1708.

Park, Katharine. 'The Rediscovery of the Clitoris: French Medicine and the Tribade 1570–1620'. *The Body in Parts: Fantasies of Corporeality in Early Modern Europe*. Ed. David Hillman and Carla Mazzio. London: Routledge, 1991. 171–93.

Paster, Gail Kern. *The Body Embarrassed: Drama and the Disciplines of Shame in Early Modern England*. Ithaca: Cornell University Press, 1993.

Planque, M. *Bibliothèque choisie de medecine tirée des ouvrages périodiques, tant français qu'étrangers [Selected library of medicine taken from periodical works, both French and foreign]*. 10 vols. Paris: veuve D'Houry, 1748–70.

Pomata, Gianna. 'Menstruating Men: Similarity and Difference of the Sexes in Early Modern Medicine'. *Generation and Degeneration: Tropes of Reproduction in Literature and History from Antiquity to Early Modern Europe*. Ed. Valeria Finucci and Kevin Brownlee. Durham: Duke University Press, 2001. 109–52.

Stolberg, Michael. 'The Monthly Malady: A History of Premenstrual Suffering'. *Medical History* 44.3 (2000): 301–22.

Tiraqueau, André. *Ex commentariis in pictionum consuetuedines sectio de Legibus connubialibus et jure maritali [Commentary on conjugal law and marital oaths]*. Lyon: G. Rouillium, 1574.

7
Menstruation and Sexual Difference in Early Modern Medicine

Michael Stolberg

Strange things happened in the bodies of early modern men. With every new moon, a young patient of Richard Mead's coughed up blood for four or five days (Mead 43). A German man experienced haemorrhoidal bleeding exactly on the eighteenth day of every month, although the months differed in length (Themmen 21). Others bled periodically from their noses, their fingertips, their penises, and every other imaginable part of the body. There are dozens of observations like these in early modern medical writing (Schurig 83–118). In this chapter, I use stories such as these as my starting point for a more general analysis of scholarly theories of menstruation and their relationship to prevailing notions of sexual difference in the period from about 1500 to 1800 CE. These stories should not be mistaken for evidence that physicians doubted that monthly bleeding commonly occurred only in women; rather, physicians collected them because they satisfied their interest in the rare, the extraordinary, and the unusual. They offer important insights into what turns out to be a rather complicated and variable relationship between two connected but nevertheless quite distinct areas of scientific discourse: theories of menstruation and theories of sexual difference, both of which were subject to much change and controversy in the early modern period. Depending on which specific theories were preferred, at a given time and by a given author or group of physicians, the link between menstruation and female biology could be framed in quite different terms. Long into the eighteenth century, theories of menstruation, like contemporary physiology and pathology in general, were almost exclusively based on a holistic, humoural understanding of the human body (Duden *passim*; Paster 64–83). Some authors underlined

90

the role of menstrual blood in human procreation, as an equivalent of the male semen or as foetal nutrition but that, role remained controversial as some women were known to have had children without ever menstruating, and menstruation was virtually unknown in species other than humans (Joubert 62; Paré 576; Stilting 1). There was no serious doubt, however, up to around 1750, about menstruation's function as a major pathway for the evacuation of superfluous or harmful matter from the whole body. We can roughly distinguish three major explanatory models within this holistic humoural model: the cathartic, the plethoric and the iatrochemical.

The cathartic, the plethoric and the iatrochemical models

According to the cathartic theory of menstruation, which dominated scholarly debates until about 1580 and informed medical lay culture far into the nineteenth century, menstruation served primarily as a means to free women from the poisonous, morbific, impure matter which constantly accumulated in the female body. The power of this poison could also be seen from its deadly effects on plants and insects, and the mere presence of a menstruating woman was enough to blind mirrors, to spoil wine or salted meat and to cause abortion and infertility (Fernel 321; Fuchs 200–3; Rocheus 133–47). No wonder then that a 'suppressed' menstruation or even just natural menopause could work the most pernicious effects on a body swamped with poisonous matter. Hysteria or 'uterine suffocation', the result of poisonous vapours, and scirrhus or cancer, developing from local deposits of menstrual flux, were only the two most conspicuous among the many potential consequences (Fernel 321; Fuchs 200–3; Marinello 87^v). Men might sometimes also accumulate impure, morbific matter in their bodies and they, too, might get rid of impure peccant matter at more or less regular intervals, through, for example, haemorrhoidal bleeding. But healthy men did not need this kind of evacuation, whereas women did because they were naturally cooler and moister than men and, in particular, their vital, 'innate heat' was weaker (Marinello 86^v–87^v; Massaria 483–95). They were not as well equipped as men to 'cook' or 'concoct' the food to the degree which was necessary to completely assimilate it and to dispel the impure remainders through the skin, as sweat or body hair. Female menstruation thus bordered on the pathological from the perspective of the 'standard' male body. It was, in contemporary terms, the 'female disease', though, to be precise, the need for menstruation, not the evacuation itself, was pathological. The evacuation was, under the given circumstances of female imperfection, very welcome.

Some historians have mistaken the cathartic model of menstruation for the dominant early modern theory of menstruation. But already by 1600 this 'pathological' view of menstruation was a minority position amongst learned physicians (Pomata 64–83; Stolberg, 'Erfahrungen' 918–21). It had been largely replaced by the new 'plethora model' of menstruation which dominated scholarly debates until the last decades of the eighteenth century. The allegedly harmful, poisonous effects of menstrual blood on plants, animals and men were now widely rejected as fictitious. In healthy women, the menstrual blood was 'as pure as any other blood in the body' (Laurens 602–6; Platter 525; Ross 87 (qtd.); Varandaeus 492) and some authors even praised its pleasant fragrance (Freind 4). Occasionally, in sick women, the menstrual flux might also serve to evacuate some impure, peccant matter but that was not normally the case (Liébault 329, 337–41; Mercuriale 101). The need for menstruation followed, in this new model, exclusively from woman's capacity to accumulate more good, healthy, nutritious blood in her body than she needed and could consume (Akakia 745f; Joubert 59f). In pregnant women, this blood nourished the foetus and turned into post-partum milk, which explained why menstruation temporarily ceased during that time. Non-pregnant and non-lactating women had to get rid of the blood at regular intervals because of its sheer volume, otherwise the vessels would soon be excessively filled with blood and the natural flow of the humours disturbed. Ultimately the stagnating blood would turn bad and putrefy, as many cases of a 'suppressed menstruation' showed.

As to the driving force of menstruation, most sixteenth- and early seventeenth-century authors took the regular monthly evacuation as a striking example of the periodical activity of nature in general and the 'expulsive faculty' in particular. Along similar lines, eighteenth-century subscribers to the thinking of George Ernst Stahl attributed the periodical menstrual flux directly to the purposeful workings of soul or nature. The large majority of physicians in the later seventeenth and early eighteenth centuries, however, resorted to mechanistic, hydraulic terms rather than to Galenic faculties, nature or the soul (Bertrand *passim*; Freind *passim*; Hoffmann 223–32). Every month, they claimed, the accumulating blood filled and extended the vessels more and more. An over-abundance of blood developed, a so-called 'plethora', in the whole body and, as some authors stressed, particularly in the uterus (Astruc 14–60). For this reason, many women, at the approach of their periods, suffered from heaviness and pulling pains, especially in the lower abdomen and the legs, but also from headaches, dizziness and other symptoms then considered as typically plethoric (Astruc 17f; Stolberg, 'Monthly Malady' 305–7). Finally the vessels could no longer resist the pressure. Usually the vessels in the highly vascularised uterus with its low position in the body gave way but sometimes the blood also found its way through other pathways by way of a 'vicarious menstruation'. Once the body had acquired this 'habit', regular and sufficiently copious menstrual flux was still considered essential for female health. But menstruation was no longer the sign of a naturally inferior or indeed pathological female body. On the contrary, it was evidence of woman's indispensable, complementary role in God's creation. The visible menstrual flux was only a by-product of the unique and highly appreciated female capacity to accumulate a surplus of good, pure blood for foetal nutrition.

This reinterpretation of the nature, genesis and purpose of menstruation was closely linked to a roughly simultaneous more general change in the anatomy of sexual difference around 1600. From the sixteenth century onwards, scholarly medical theories on sexual difference definitely shifted from a model of female inferiority to a model of complementary otherness which stressed the fundamental and incommensurable difference between male and female organs and bones. Thomas Laquer argues in *Making Sex* that that early modern medical notions of sexual difference were based on a 'one-sex model' of female inferiority. Early modern physicians, according to Laqueur, used the famous Galenic analogy between men and female genitals, that is, the vagina was an inverted penis and the body of the uterus an inverted scrotum. The crucial difference was that the female genitals remained inside the body because women were less perfect than men, because they were colder and their innate heat, the decisive active principle in the body, was weaker. According to Laqueur, it was only in the *eighteenth* century that a new anatomy of incommensurable dimorphism emerged. Enlightened beliefs in universal humanity and equal rights made physicians seek naturally given, physiological differences between the sexes which still could justify the traditional, patriarchal order and women's confinement to the domestic sphere. As I have shown elsewhere, Laqueur's conclusions are seriously flawed ('A Woman' 285–9). Although he is dealing with the scientific discourse of a period in which Latin was still by far the dominant language, Laqueur relies on only a handful of vernacular texts and he makes much of anatomical illustrations which, particularly before 1650, seem to endorse a model of analogy. This is despite the well-known fact that anatomical illustrations were then often linked only very loosely to the respective texts and followed their own iconographic traditions. Already by 1600, dozens of the most influential and widely quoted physicians writing on anatomy or 'women's diseases' insisted in their (mostly Latin) works almost unanimously on the fundamental anatomical difference between the male and female genitals. Rather than regarding woman as a mere inferior, imperfect variant of man, they stressed the unique anatomical structure of the female body which they also declared to be perfect in its own right.[1] The claim that the eighteenth century grounded sexual difference for the first time in biology while the humoural model aimed only at a 'metaphysical truth' is anachronistic. Early modern medics saw and presented themselves in a very literal and explicit manner as 'physici' or 'physicians'. Their field of (increasingly empirical) enquiry and expertise was not metaphysics but the 'physica', that is, the things and laws of nature in general and in human physiology and pathology in particular.

The major late seventeenth- and early eighteenth-century alternative to the plethora model of menstruation was the iatrochemical model (Charleton 90–108; Ettmüller 69; Graaf 135f.; Mauriceau 44–8). Iatrochemists rejected the plethora model largely on the very grounds just discussed: since plethora frequently occurred in men as well, a mere abundance of blood in the female body could not, in their view, account for menstruation. Instead, many iatrochemists relied on the notion of a specific menstrual 'ferment' which drove the blood or humours into intense commotion every month, stretching and expanding them to their utmost limits. As in the plethora model, the vessels gave way at the point of least resistance, which was usually the uterus. The allegedly 'typical' menstrual complaints once more served to support menstrual theory. Throughout their periods – and not just at their approach – many women, it was argued, suffered from pain in their belly and head as well as from sensations of heat or fire around their genitals and from sudden, hot, feverish waves rising in their bodies, all of which was taken as evidence for the fermentative commotion (Charleton 93; Duncan, preface; Ettmüller 69f; Graaf 132; Musitano 56; Stolberg, 'Monthly Malady' 307–9). The principle purpose of menstruation was, according to most proponents of this model, to free the body of excremental impurities, just as wine and beer fermentation separated out the gross matter (Motte 75f; Musitano 55). In this respect, most iatrochemical models provided a modernised version of the cathartic model although providing a more negative view of menstrual blood than the plethora concept. Female genitals were likened to a 'gutter', it was asserted that menstrual blood had deleterious effects on plants and animals and that the presence of menstruating women disturbed the natural fermentation of wine or bread (Charleton 91; Motte 72f; Musitano 47). The teleological interpretation of the monthly rhythm ran along similar lines: if women's impurities flowed constantly from her body, rather than on specific days of the month, men would loathe any sexual intercourse with women and this would jeopardise the future of mankind (Stilting 2). Opinions about the exact nature and genesis of this fermentative or chemical process varied. Some authors felt that the mere presence of impure, foreign matter in the blood was sufficient to cause fermentation or effervescence, which could, in principle, also happen in men. Leading proponents of the iatrochemical model, however, linked menstruation to a specific female menstrual ferment or even framed menstruation literally as 'Eve's curse'. The menstrual ferment, they claimed, had entered Eve's body when she ate the forbidden apple in paradise and this was passed from mother to daughter, as a constant reminder of female sinfulness (Duncan, preface 67; Helmont 448; Musitano 47–52).

Male menstruation and the standard body

The plethora model provided an unprecedented opportunity for identifying and accepting 'menstrual' bleeding in men as well. In contrast to the accumulation of impure, excremental matter, the ability to produce good, nutritious blood was in both sexes indispensable for health. In certain ways, men were even better equipped for this. Their stronger heat could 'concoct' the food more perfectly. Numerous medical case reports proved that men who ate well and moved little tended to accumulate blood in their bodies. Rather than just having a healthy sanguine constitution, they risked becoming outright plethoric. They suffered from headaches and sensations of fullness and tension in their whole body and might ultimately succumb to apoplexy. Fortunately, the surplus blood was often eliminated via bleeding from the nose, the haemorrhoids or other parts of the body (Rivière 93f). Medical practice could support this through blood-letting or cupping. However, when the blood accumulated at a fairly steady rate the evacuation had to take place periodically, at more or less regular intervals. The plethora model of menstruation challenged physicians not so much to explain why periodical bleeding could also occur in healthy men but why it was nevertheless largely confined to women.

There were two major approaches to this issue. Firstly, women's heat was understood to be strong enough to produce large amounts of blood from food but too weak to 'cook' and thereby assimilate all the blood in the third and final step of concoction (Mercuriale 99). On the basis of this explanation, menstruation would have been an exclusively female event only if the strongest female heat was still weaker than the weakest male heat – a notion already disputed by the sixteenth century. Indeed, some authors even doubted that women were generally colder at all. Women's heat, they argued, was only less perceptible because of their greater moistness (Akakia 745f; Scaliger 832–8). The second approach drew upon the traditional Galenic notion that women consumed less blood because they lacked physical exercise and were generally confined to a more domestic, sedentary life (Fuchs 498; Gordon 617). Some eighteenth-century medical critics of the unhealthy effects of modern, urban life-style expanded on this notion. In their view, menstruation (and menstrual complaints) resulted primarily from the idle life-style and excessively rich nutrition of aristocratic women, whereas peasant women menstruated less copiously and some women – for example those in Greenland or Brazil – hardly needed menstruation (Astruc 59f; Roussel 195f). This left unclear, however, why very few affluent, well-nourished and idle gentlemen periodically evacuated blood, while most hard-working women from the lower classes did. Both approaches thus could not explain periodical bleeding as a uniquely female phenomenon. The plethora model made periodical bleeding in men as plausible and acceptable as a 'normal' event. The two roughly simultaneous shifts, from catharsis to plethora in menstrual theory and from a primarily humoural theory of sexual difference to one which focussed on solid, anatomical, structural difference, put the relationship between the two on new grounds. While humours, qualities and innate heat were still used to explain menstruation, they no longer carried the burden of securing a clear-cut line between the sexes. This task had been taken over by sexual anatomy. Far from providing evidence for a predominant one-sex model, stories of 'menstruating men' were, on the contrary, fostered by a wide-spread insistence on organic anatomical difference which made such stories unproblematic, because a clear-cut demarcation between the sexes could nevertheless be maintained.

In her ground-breaking study of early modern cases of 'male menstruation', Gianna Pomata has not only shown how these cases made sense within the explanatory framework, described above as the plethora model, but also used them to reject Laqueur's claim that early modern medics always saw the male body as the standard body and the female body as deviant. Accepting Laqueur's framework of a predominant 'one-sex model' in early modern medicine, she has argued that in the case of 'male menstruation' it was clearly the female rather than the male body which provided the measure of comparison. Pomata might have found even better evidence for her argument in contemporary writing about the male breasts, which some authors frankly described as 'imperfect' (Laurens 723). However, for various reasons, reports about 'male menstruation' cannot be taken to imply that female menstruation was the standard. Firstly, 'menstrual' or 'menstrualis', as early modern physicians were well aware, was derived from the Latin 'mens' and meant 'monthly'. A 'menstrual haemorrhagia' in a male patient was quite simply a bleeding which occurred every month and did not automatically imply comparison with female menstruation. Also, notions of rhythmical, periodical change generally played an important role in early modern medicine, in men and women alike. Moreover, the older theory of critical days remained one cornerstone of learned medicine and is of immediate relevance to the context of menstruation. This theory postulated that the course of diseases changed for better or worse on given days of the disease. Menstruation derived much of its contemporary meaning from this framework and was sometimes explicitly described as a 'critical' evacuation or 'critical' time (Roussel 183; Stahl *passim*). In this sense, the critical evacuation of matter at monthly intervals in the form of female menstruation was only a particularly striking case among a wide range of similar critical evacuations in both sexes. Along similar lines, Santorio Santorio's widely quoted findings were that men gained between one and two pounds every month which they got rid of, at monthly intervals, by a 'critical evacuation of turbid urine' (19). While it is tempting to interpret Santorio's findings as suggesting another kind of male 'menstruation', within the framework of contemporary humouralism such evacuations referred to the fluid economy of the body in general, which, in the eyes of contemporary physicians, was, if anything, male. Thus it is clear that there is a substantial risk of anachronistic misinterpretation, if we use the modern notion of periodical bleeding as a token of femininity in order to make sense of early modern accounts of monthly or 'menstrual' bleeding.

Indeed, if we look at the various theories, one could argue as well that female menstruation derived its medical meanings from comparison with the male body as the implicit standard and norm of human biology. The cathartic and iatrochemical models interpreted the need for a periodical evacuation as essentially pathological, as a deviation from the healthy male body which did not accumulate impure, peccant matter in the first place. The plethora model of menstruation, in turn, built on more general notions of the effects of over-abundant blood in the body – and above all in the male body, because plethora was much more commonly observed in men. Seen from this perspective, female menstruation was just a special, particularly regular, evacuation amongst a wide range of similar means by which male and female bodies got rid of superfluous blood and other humours, at more or less periodical intervals.

Vitalism

While iatrochemistry and the notions of a menstrual ferment rapidly lost ground in eighteenth-century medicine, the plethora theory of menstruation gained renewed vigour for some time, especially in the wake of John Freind's widely quoted 'Emmenologia' (1703). Freind offered the most complete and systematic account of the plethora theory to date and used it, in turn, to demonstrate the value of a hydraulic understanding of human physiology in general. In the second half of the eighteenth century, however, the humoural hydraulic paradigm, and the plethora model of menstruation with it, was, in turn, increasingly replaced by new theories such as vitalism, which which focused on the specific vital faculties and functions inherent in individual organs, tissues or fibres rather than on overall humoural physiology. The vitalists' paradigm was 'glandular secretion'. Menstruation was no longer explained in hydraulic terms, of passive filtration under pressure, but conceived of as a process of active preparation and secretion of specific substances, by organs or tissues specifically endowed for that purpose. The new model was directly applied to the interpretation of menstruation, though in various forms. At first, menstruation was reframed primarily as a specific, active secretion of the uterus (Barthez; Bordeu 191; Gardien vol. I, 222; Salmon 3; Vigarous 71; cf. Müller-Hess *passim*). This fitted well with the role, which many late eighteenth- and early nineteenth-century physicians assigned the uterus of the dominant, driving force in the female body. As a result of its power over the body, menstruation as an active uterine secretion was now described, above all, as a massive irritation of the body and the nervous system in particular. Many women were said to suffer intensely during their periods, in particular from nervous symptoms. The physiological role of menstruation, on the other hand, remained controversial and receded into the background (Roussel 207). The traditional overwhelmingly negative interpretation of menopause was increasingly replaced by the view that women, once they had survived the somewhat troubled 'change of life', often enjoyed better health and lived longer than men (Gardien vol. I, 367; cf. Stolberg, 'Woman's Hell' 416f).

Over the course of the nineteenth century, the focus shifted from the uterus to the specific vital properties and periodical changes of the ovaries (Gendrin vol. II, 17–36). Long before the advent of sexual endocrinology, various eighteenth-century authors had already assumed the production of some specific subtle matter in the ovaries responsible for the typical 'feminine' features of women's bodies. Their claim was modelled on the widely-accepted corresponding notion of subtle testicular matter or seminal evaporations thought to produce the masculine features found in normal men and lacking in castrates (Stilting 5–7; Themmen 31–5). Nineteenth-century writers increasingly perceived the ovaries even as the decisive driving force behind menstruation and as the dominant organs in the female body as such. The various new uterine and ovarian theories shared a common assumption: menstruation was the result of a peculiar activity of specific, solid organs – the female genitals – and only indirectly linked to the humoural economy of the whole body. Mainstream menstrual theory had thus finally followed the much earlier shift in medical notions of sexual difference, from a holistic model based predominantly on fluids to a solidist model which focused on the organs and fibres with their specific activities or functions. It was only with the greatest of difficulty that 'vicarious menstruation', let alone periodical evacuations in men, could be attributed to the specific vital faculties of the female genitals. They had to be interpreted in entirely different terms. Menstruation was the exclusive privilege of women. It was an outstanding token of femininity and sexual difference. After some 300 years of temporary divergence, the interpretive frameworks of menstrual theory and sexual difference had been realigned.

Note

1. Laqueur has more recently conceded that at least some pre-eighteenth-century physicians subscribed to dimorphism, but he disavows that the overwhelming majority of extant treatises do not support his view and still insists that this had 'minimal impact' because the one-sex model was so firmly entrenched ('Sex' 300).

Works cited

Akakia, Martin. 'De morbis muliebribus'. *Gynaeciorum sive de mulierum tum communibus, tum gravidarum, parientium, et puerperarum affectibus et morbis libri.* Ed. Israel Spachius. Strasbourg, 1597. 745–801.
Astruc, Jean. *Traité des maladies des femmes.* Lyons, 1765.
Barthez, Paul Joseph. *Nouveaux éléments de la science de l'homme.* Montpellier, 1778.
Bertrand, Thomas Bernard. *Quaestio medica: An catamenia a plethora?* Paris, 1711.

Bordeu, Théophile. 'Recherches anatomiques sur la position des glandes, et sur leur action'. *Oeuvres complètes*. Vol. 1. 1752. Paris, 1818. 45–208.

Charleton, Walter. *Inquisitio physica de causis catameniorum et uteri rheumatismo.* London, 1685.

Duden, Barbara. *The Woman Beneath the Skin: A Doctor's Patients in Seventeenth-century Germany.* Cambridge: Harvard University Press, 1991.

Duncan, Daniel. *Seconde partie de la chymie naturelle ou l'explication chymique et méchanique de l'évacuation particulière aux femmes.* Paris, 1687.

Ettmüller, Michael. *Pratique de médecine spéciale sur les maladies propres des hommes, des femmes et des petits enfans.* Lyons, 1691.

Fernel, Jean. *Universa medicina.* 1543. Leiden, 1645.

Freind, John. *Emmenologia.* 1703. London, 1729.

Fuchs, Leonhart. *De sanandis totius humani corporis malis libri quinque.* Lyon, 1547.

Gardien, Claude-Martin. *Traité complet d'accouchemens, et des maladies des filles, des femmes et des enfants.* 3rd ed. Paris, 1824.

Gendrin, A.N. *Traité philosophique de médecine pratique.* Paris, 1839.

Gordon, Bernard. *Lilium medicinae.* Lyons, 1550.

Graaf, Regnier de. *De mulierum organis generationi inservientibus tractatus novus.* Leiden, 1672.

Guyon, Loys. *Le miroir de la beauté et santé corporelle.* Lyons, 1625.

Helmont, Johann Baptist van. *Ortus medicinae.* 4th ed. Lyons, 1655.

Hoffmann, Friedrich. *Medicina rationalis.* Frankfurt, 1738.

Joubert, Laurent. *La première et seconde partie des erreurs populaires, touchant la médecine et le régime de santé.* Rouen, 1601.

Laqueur, Thomas W. *Making Sex. Body and Gender from the Greeks to Freud.* Cambridge: Harvard University Press, 1990.

——. 'Sex in the flesh'. *Isis* 94 (2003): 300–6.

Laurens, André du. *Historia anatomica humani corporis partes singulas vberrime enodans.* Frankfurt, 1602.

Liébault, Jean. *Les trois livres de la santé, foecondité, et maladies des femmes.* Paris, 1582.

Marinello, Giovanni. *Le medicine partenenti alle infermità delle donne.* Venice, 1563.

Massaria, Alessandro. *Practica medica.* Lyons, 1616.

Mauriceau, François. *Traité des maladies des femmes grosses, et de celles qui sont accouchées.* 3rd ed. Paris, 1681.

Mead, Richard. *Of the Power and Influence of the Sun and Moon on Humane Bodies.* London, 1712.

Medicus, Casimir. *Traité des maladies périodiques.* Paris, 1790.

Mercado, Luìs. *De mulierum affectionibus libri IIII.* Venice, 1587.

Mercuriale, Girolamo. 'De morbis muliebribus'. *Gynaeciorum sive de mulierum tum communibus, tum gravidarum, parientium, et puerperarum affectibus et morbis libri.* Ed. Israel Spachius. Strasbourg, 1597. 7–195.

Messerus, Joh. Fridericus. *De naturali mensium suppressione.* Giessen, 1711.

de la Motte, Guillaume. *Traité complet des accouchemens naturels, non naturels, et contre nature.* Paris, 1702.

Müller-Hess, Hans Georg. *Die Lehre von der Menstruation vom Beginn der Neuzeit bis zur Begründung der Zellenlehre.* Berlin, 1938.

Musitano, Carlo. *De morbis mulierum tractatus.* Coligny, 1709.

Paré, Ambroise. *The Works.* London, 1678.

Paster, Gail Kern. *The Body Embarrassed: Drama and the Disciplines of Shame in Early Modern England*. Ithaca: Cornell University Press, 1993.

Platter, Felix. *Praxeos medicae opus*. 3rd ed. Basel, 1666.

Pomata, Gianna. 'Uomini mestruanti. Somiglianza e differenza fra i sessi in Europa in età moderna'. *Quaderni storici* 79 (1992): 51–103.

Rivière, Lazare. *Opera medica universa*. Lyon, 1698.

Rocheus, Nicolaus. 'De morbis mulierum curandis liber'. *Gynaeciorum sive de mulierum tum communibus, tum gravidarum, parientium, et puerperarum affectibus et morbis libri*. Ed. Israel Spachius. Strasbourg, 1597. 128–221.

Ross, Alexander. *Arcana microcosmi: or, the hid secrets of man's body disclosed*. London, 1651.

Roussel, Pierre. *Systeme physique et moral de la femme*. Paris, 1775.

Salmon, Nicolaus. *Dissertatio physiologica de fluxu menstruo*. Montpellier, 1745.

Santorio, Santorio. *De statica medicina*. The Hague, 1657.

Scaliger, Julius Caesar. *Exotericarum exercitationum Libri XV de subtilitate ad Hieronymum Cardanum*. Frankfurt, 1576.

Schurig, Martin. *Parthenologia historico-medica*. Dresden and Leipzig, 1729.

Stahl, Georg Ernst. *Proempticon inaugurale de fluxus muliebris, quatenus menstrui causa*. Halle and Magdeburg, 1702.

Stilting, Aegidius. *Disputatio medica de fluxu menstruo*. Praes. Joannis Munnicks. Utrecht, 1708.

Stolberg, Michael. 'Erfahrungen und Deutungen der weiblichen Monatsblutung in der Frühen Neuzeit'. *Scientiae et artes. Die Vermittlung alten und neuen Wissens in Literatur, Kunst und Musik*. Ed. Anselm Steiger. Wolfenbüttel: Barbara Mahlmann-Bauer, 2004. 913–31.

——. *Homo patiens. Krankheits- und Körpererfahrung in der Frühen Neuzeit*. Cologne: Böhlau, 2003.

——. 'The Monthly Malady: A History of Pre-Menstrual Suffering'. *Medical History* 44 (2000): 301–22.

——. 'A Woman Down to her Bones: The Anatomy of Sexual Difference in Early Modern Europe'. *Isis* 94 (2003): 274–99.

——. 'A Woman's Hell? Medical Perceptions of Menopause in Early Modern Europe'. *Bulletin of the History of Medicine* 73 (1999): 408–28.

Themmen, Phoebus Hitzerus. *Dissertatio physiologico-medica inauguralis de mensibus, ex materia quadam peculiari, ovariis secreta, oriundis*. Leiden, 1781.

Varandaeus, Joannis. *Opera omnia theorica et practica*. Basel, 1658.

Vigarous, Joseph-Marie-Joachim. *Cours élémentaire de maladies des femmes, ou essai sur une nouvelle méthode pour étudier et pour classer les maladies de ce sexe*. Paris, 1801.

8

'I Believe It To Be a Case Depending on Menstruation': Madness and Menstrual Taboo in British Medical Practice, c. 1840–1930

Julie-Marie Strange

On 14 June 1856 a twenty-four-year-old woman, Sarah Ann, was found hanging in the water closet of Horton Road Asylum in Gloucester. Suffering from severe despondency since the miscarriage of a child some months previously, Sarah Ann had twisted a menstrual napkin around her throat in a bid to destroy herself.[1] Whether Sarah Ann was menstruating or not at the time of this suicide attempt is unclear. However, her use of a menstrual napkin as a device for self-destruction provides an arresting, if dramatic, metaphor for the experience of menstruation as defined and understood by medical practitioners in Victorian Britain: women would never enjoy equality with men because they were stifled and suffocated by the inherent sickness of their own biology.[2] Within a model of increasing professionalisation, nineteenth-century medical discourse maintained that it represented the world accurately and used its empirical claims to assert that science had proven, and could even improve, the laws of nature. However, medical claims are inextricable from their cultural contexts and paradigms of menstruation were created, understood and interpreted in direct relation to perceptions of femininity. This chapter demonstrates that medical and cultural beliefs are dialectical, each sphere informing the other whilst operating to maintain a sex-based status quo. Notably, medical discourse rendered menstruation a problematic physiological function whilst creating (and perpetuating) a conception of women that implicitly privileged marriage and motherhood; periodically sick and subject to emotional instability, women could not expect to compete in the public world on an equal footing with men. Menstruation was an illness of femininity, yet paradoxically, it was an illness that was essential to conceptions of female health and, more importantly, fertility. Of course, pregnancy was the ultimate cure for this illness. As late as 1926, the authoritative medical journal the *Lancet* published an article entitled 'Menstruation and Pain' which seemed to suggest that if women behaved as nature intended, continual pregnancy and lactation would render menstruation almost obsolete (Anon., 'Menstruation' 611–12). Paying particular attention to the relationship between menstruation and mental health, this chapter explores medical paradigms of menstruation in the nineteenth and early twentieth centuries to demonstrate the intersections between medicine and culture. In particular, this chapter engages with conceptions of menstrual 'madness' to suggest that perceived relationships between the menstrual cycle and mental health were shaped by a desire to regulate feminine behaviour and the periodic flow. Placing such an understanding within a broader context, this chapter concludes with an analysis of shifting paradigms of menstruation in the early twentieth century to highlight the inter-relationship between medical narratives of menstruation and changes in women's status in society.

Sexual knowledge and power

As branches of specialist inquiry and practice, gynaecology and obstetrics mushroomed from the late 1830s and the topic of menstruation featured with increasing regularity in medical journals such as the *Lancet* and the *British Medical Journal* under the generic heading 'Diseases of Women'. Despite all the interest in menstruation, practitioners often described the subject as unfortunate, unpleasant, distasteful, troublesome, and tedious, and they remained confused about the role of menstruation, and its precise relationship with ovulation, until the late 1920s.[3] The relationship between practitioner and patient was typified by an imbalance of power. Sexual knowledge, even when it concerned the female body, was considered too shocking for the modesty of women. Thus, gynaecology was limited to the expertise of men who asserted their status as the protectors of women. Reviewing a popular advice text, the *Complete Ladies' Guide to Physiology*, in 1854, the *Lancet* exclaimed in horror that the ladies of England had been presented with illustrations of vaginas, the uterus, childbirth and even semen. That it had been written by a fellow medical man who 'should help to preserve the young innocent and undefiled' only exacerbated the insult to women and the profession: 'We say advisedly that such rubbish has never before been put into type; and we blush to think, indeed we feel deeply humiliated that the author is a brother practitioner' (Anon., Review of *Complete* 103).

The desire to preserve physiological knowledge within the domain of the male medical expert highlights just one facet of the power imbalance that typified gynaecological encounters. Whilst the manifest purpose of gynaecology was to nurture the 'feminine' generative capacity, women were often sexualised in medical texts in a language of danger, violence and immorality (Lansbury 413–27). Case reports might refer to thrusting into the vagina, plunging through a hymen, holding women still or 'keeping the woman quiet' (Tanner 363–4). Controversies over the internal examination of women, especially with the increasingly sophisticated speculum, make even more explicit the relationship between power and biology. Although some practitioners expressed uncertainty with regard to conducting gynaecological examinations on virgins, or worried about offending the modesty of women, other experts advised that it was foolish to allow embarrassment to hamper treatment, not least because the reputation of the gynaecological expert depended upon pioneering and authoritative research (Usher-Somes 1010–11). The notorious American practitioner J. Marion Sims was quoted in the *Lancet* advising that the best patients to operate on were 'delicate, timid women, who were conscious that something was being done' but not sure what (224–5). Other practitioners, however, feared that the gynaecological encounter would unleash base sexual passions in patients, situating the practitioner as an object of lust and encouraging women to fabricate or exaggerate complaints to satisfy their sexual desires. At least one distraught husband cited the gynaecological examination as the cause of breakdown in his marriage.[4] Hence, whilst women were the unwitting victims of their own bodies, men were also liable to be represented as victims of the unchecked female body.

Along with puberty, pregnancy, childbirth, and the menopause, menstruation represented a 'sexual crisis', that is, a period of genital and reproductive excitement which intensified the inherent instability of the 'feminine' body. Puberty and menopause were referred to respectively as the 'advent' and 'decline' of femininity, thus rooting womanhood firmly in relation to fertility (Tyler-Smith, 2 Feb. 1856 113–14). Notably, the menopausal woman was compared to 'those animals who die when once they have transmitted life to others' (Tilt, 1855 564–6). As Mary Poovey has observed, Victorian medical narratives dissolved femininity into one 'enormous universal uterus'; furthermore, against the stable norm of 'man', the universal uterus was perceived as pathological, uncontrollable and excitable (35). Integral to the juxtaposition between the stable male and the unstable female was the belief that male sexual emissions were healthy and fundamental to sustaining masculine vitality. In comparison, female sexual processes were overwhelmingly discussed in negative language that fostered associations between illness and the female reproductive organs. The assumption that menstruation was debilitating was widely held. Dysmenorrhoea, a loosely defined category for describing painful periods, was seen for much of the nineteenth century as a normative experience. Conceptions of menstruation as pathological are reflected in the evocative terms used to discuss it. In a series of lectures on the 'Diseases of Women' in 1853, the pioneering specialist William Tyler-Smith referred to menstruation variously as a 'monthly illness', 'catamenial derangement', 'decline' and 'periodical flow' (509–10). Another renowned expert, Robert Barnes, described it in 1880 – again in lectures on the 'Diseases of Women' – in terms of decay and dissipation: menstruation was a 'missed pregnancy', an abortion, the 'breaking down and exfoliation of the decidua' (121–3). Other practitioners referred to it as being 'unwell', a time of 'crisis' and, curiously, the 'dodging time'. Conversely, semen tended to be described in a language of vitality. As social anthropologist Emily Martin notes of contemporary medical models of reproduction, menstruation is the wrong ending to a fairytale where happiness and resolution hinge on the gallant sperm charging up the vagina to rescue the femme fatale egg from oblivion (104).

Menstrual instability

In light of such paradigms, it was unthinkable that women could participate in society on an equal footing with men. By virtue of biology, not culture, women were best suited to domesticity. As the eminent mental-health specialist Henry Maudsley put it in 1874, women had to 'bear up against the physical deterioration' of their monthly periods (663–4). In particular, menstruation heightened mental and emotional volatility by inducing erratic impulses which weakened self-control and distorted mental vision. Most practitioners subscribed to a physiological paradigm whereby the female body contained a fixed amount of energy. Menstruation represented a threat as it necessitated a diversion of energy to the reproductive organs and away from the brain and thus hysteria was identified as 'one of the most well known cerebral symptoms of menstruation' (qtd. Anon., Review of 'On the Diseases' 270–1). Tyler-Smith considered an increase in irritability and predisposition to hysteria a common experience in the days prior to menstruation (2 Feb. 1856 113–15). Francis Anstie concluded in 1872 that hysteria was 'but the exaggeration of features which are common to the feminine characteristics' thus demonstrating the extent to which womanhood was perceived as inherently unstable (839–42). Barnes perpetuated the assumption that menstruation exposed the vulnerability of female mental health: the main characteristic of menstruation was 'exalted nerve tension, the emotional element is often predominant; the temper is more sensitive'. It followed, according to Barnes, that since many cases of hysteria, insanity or depression might well be located in the menstrual cycle, women committed to lunatic asylums should have immediate access to a gynaecologist (1867 530; 1896 1797; cf. Macnaughton Jones, 1900 446). In Barnes' rhetoric, the gynaecologist was cast in the role of knight in shining armour, an image reinforced by his colleague Amand Routh who maintained that unless placed in 'good hands' (those of the gynaecologist), countless women would drift needlessly into asylums (1896 1797).

Collaboration between the gynaecologist and the 'alienist' (or 'head doctor') was epitomised in the careers of prominent psychiatrists such as George Savage and Henry Maudsley. Yet, at a provincial level, the perceived relationship between madness and menstruation appeared to provide a framework for thinking about and dealing with mental illness. Focusing on the case records of two lunatic asylums, Horton Road Asylum in Gloucester, where Sarah Ann had tried to hang herself with a menstrual napkin, and Pen-y-Fal in Abergavenny, South Wales, it is possible to explore the ways in which practitioner narratives positioned menstrual health within a correlative relationship with insanity and/or emotional excitement. Compared to hereditary predisposition, poverty, grief, disappointment in love, and 'profligate living', menstruation was rarely cited as a direct cause of insanity. Rather, it tended to be represented as inseparable from factors associated with the mental state of a patient, both in terms of decline and recovery. Notably, on entry to an asylum, all patients were given a physical examination. For women, this often included a gynaecological examination. In addition, female patients and/or their kin were asked to relate a menstrual history, paying attention to the age when menstruation began, the regularity of periods, the duration of the period and the quantity of blood lost. They were also asked to describe the physical and emotional effects of menstruation.

Details of menstruation were, of course, important for establishing that the patient was not pregnant. Yet these details were also used to determine and describe the cause and the character of their mental illness.

Rhanna B., aged twenty-five, was admitted to Pen-y-Fal Asylum in 1870 with her first attack of insanity. The admission notes describe a stout young woman in moderate bodily health. The symptoms of her condition were detailed thus: a dislike of her brother and father, an inclination to rise earlier than necessary in the mornings, a tendency to mutter to herself and to refuse to undress in the evenings. Her admission notes concluded that '[t]he girls [sic] behaviour is very listless and sullen, hardly easy to move about except from one room to another. [A]nswers questions reservedly, appears to me to have something troubling her. I believe it to be a case depending upon menstrual irregularity'.[5] One week following admission, Rhanna continued to refuse food and remained sullen. Her notes report, however, that menstruation was successfully established and her mental condition had not deteriorated any further, thereby implicitly linking the two.[6] When twenty-eight-year-old Martha M. entered Pen-y-Fal, she had not menstruated for more than a year. Previous to that, her periods were described as irregular and scanty. Noting that Martha's mother was 'always very simple minded', the admission record cites amenorrhoea (suppression of the menses) as the direct cause of insanity.[7] Frequently, the rationale informing the role of suppressed, irregular or scanty periods in the decline of mental health is unclear. The admission details of Eliza E. note that 'after [finding a mutilated body in a ditch] she had numerous hysterical fits for many months and has ever since been weakly and hysterical. Menstruation being deficient she has been in a state of hysterical mania for a week'. The construction of this admission's entry thus positions the trauma of finding a mutilated body as secondary to the absence of menstruation in precipitating hysteria, despite common theories that shock was a principal factor in arresting the menses.[8]

Close monitoring of menstrual regularity and loss throughout the patient's stay in the asylum permitted practitioners to interweave accounts of menstrual health with narratives of recovery or deterioration. Thus, if the absence of menstruation could diminish one's mental state, it followed that re-establishment of the monthly courses could aid recovery. Rosa W., aged 17, was admitted to Pen-y-Fal in 1886, her case notes stating that she had a hereditary disposition to insanity, was wild in her demeanour and talked in a manic and incoherent manner. She also suffered from amenorrhoea. Having been prescribed Blaud's (iron) pills to restore the menses, her case notes state some six months later: 'Has menstruated since date of last entry and is considerably improved mentally though evidently naturally feeble of mind.' She was discharged one month later.[9] Jane P. made a similarly rapid recovery from acute mania: 'Menstrual functions have been re-established and she is rather better mentally'.[10] At Horton Road Asylum, Jane M., aged twenty-five, was 'very much quieter and rational since menses began to flow'.[11] Margaret S., a domestic servant, had been depressed and low for six weeks and had not menstruated for five or six months. The physician speculated that she was 'in the family way' but examination indicated that this was not so. Treated with aloes and iron, Margaret began to menstruate some months later. From being 'flighty and excitable' only one month previously, she was now described as 'improving rapidly in mind'.[12] Some case records appear to place a disproportionate emphasis on the re-establishment of menstruation to the point of ignoring the persistence of other symptoms. Fanny J., a twenty-year-old domestic servant, had become despondent and suicidal three months before her admission to Pen-y-Fal. Her condition was linked directly to amenorrhoea. Four months following admission, her case notes state: 'Has menstruated within the last few days, during the past week has had several hysterical fits – appears to be well in mind.' Two months later, her notes concluded that 'menstruation was well established – is not far from well, has had several more hysterical fits'. The continuation of fits was clearly not incompatible with her menstrual recovery and Fanny was discharged.[13]

Models that established correlations between amenorrhoea and insanity – explicitly referred to as 'amenorrhoeal insanity' – varied. One model contended that menstruation was typified by emotional excitement and/or nervous tension. Thus, amenorrhoea led to a culmination of tension or vicarious menstruation, commonly manifest in nosebleeds but potentially manifest as madness: the re-establishment (or release) of menstruation restored the balance in body and mind. Alternatively, menstrual blood itself might be perceived as a source of toxicity which, trapped inside the body, would have adverse effects on the woman concerned. Seemingly simplistic correlations may, however, conceal a variety of symptoms as well as diverse approaches to cure. The treatment of physical complaints such as amenorrhoea (another example would be constipation) was often the only treatment which was recorded in case records, perhaps because it was something tangible and measurable. Less controllable, however, were cases where menstruation was regular but where the onset of the flow was thought to provoke or exacerbate mental or emotional instability. Mary K's bouts of mania lasted an average of ten days and were directly associated with her menstrual periods. At this time she became 'quite maniacal, shouting and praying all night, apparently in much fear and doubt'.[14] Margaret J., a twenty-one-year-old servant admitted to Pen-y-Fal, was 'always much worse during and after a menstrual period – at these times she becomes very irritable and excited and is violent and threatening to those about her ... something comes over her at times and at such times she feels as if she would like to kill someone'.[15] Susanna B., a dressmaker, admitted to Horton Road, it was noted, 'usually conducts herself well except during the menstrual period when she becomes very excited and at that time is often very spiteful and wayward and has several fits'.[16] Where aggressive, assertive and generally unfeminine behaviour by typically quiet women may have left family and friends at a loss to understand them or their complaint, menstruation could act as a metaphorical hook on which to hang uncharacteristic female traits, especially if menstruation was viewed, as folklore implied, as mysterious and powerful, or, as medical theories were apt to note, something 'primitive'. Indeed, it is worth noting that until the late 1920s menstruation was perceived as analogous to oestrus in mammals, implicitly positioning women lower down the evolutionary scale to men.[17]

Occasionally, the onset of first menstruation was cited as the cause of insanity. According to her parents, Margaret J., a seventeen-year-old servant, had always been 'a little peculiar since her first menstrual period at the age of 13'. She had become much worse in the previous twelve months however: for four days preceding each period she became 'extremely irritable and quarrelsome, and has occasionally threatened to harm her mother'. Between menstrual periods she was 'low and desponding', and threatened to drown herself.[18] When Mary H., aged twenty-three, was admitted to Pen-y-Fal, her kin told the physician that she had been 'very peculiar ever since and more especially at her menstrual periods'.[19] The father of Elizabeth W. stated that his daughter had suffered from suppression of the menses for almost twelve months and speculated that this had altered her manner. She was, he thought, much worse at the times when she should have been unwell.[20] It would be easy to demonise the asylum physician and/or gynaecologist for making apparently crude connections about menstruation and madness. Yet patients and kin also established relationships between menstruation and mental illness. This may have been at the prompting of asylum staff, but it is impossible to tell either way, not least because the understandings of the layperson could be formulated through a series of encounters. The mother of Catherine R., admitted to Pen-y-Fal asylum in 1866, gave an account of her daughter's illness in a letter sent to asylum staff that also voiced concern that Catherine appeared to be making little progress. The letter stated that Catherine had caught whooping cough at 'a very delicate time for young ladies'. Forthwith, she 'was not regular. What ought to come away do clot in her womb and it rise to the brain'. Catherine would not 'come well until that ran from there'. The explanation had, the author noted, been formulated after ruminating over various medical, clerical and lay opinions communicated to her over some time.[21]

There was also a preponderance of women aged over forty in both asylums whose insanity was located in the 'Change of Life' and cessation of woman's reproductive purpose. This is perhaps unsurprising. Like menstruation, the effects of the menopause were manifest in physical and mental symptoms, yet for many practitioners the menopause precipitated a crisis in femininity itself: with the end of fertility, 'womanhood' lost its fundamental purpose. According to the respected practitioner George Savage in 1903, the crisis of menopause was far more dangerous if older women were inclined to introspection, asking what their role in society was and, consequently, whether men would find them attractive (1209–13). Savage was particularly adept at creating lurid images of women at the mercy of their reproductive organs. Curiously, the principal victims in his narratives were harangued husbands, vicars and medical men who became the object of the menopausal woman's attentions. Unleashed from the role of procreator, post-menopausal women were apt to become lusty and lose their sexual modesty, attributes which at least Savage found terrifying.

Overall, what emerges from the case notes of female asylum patients is a desire for menstrual experience to be regulated in terms of quantity, periodicity and quality. Menstrual flow was continually being measured against an ideal and abstract concept of the universal menstrual cycle. Thus, periods were described as 'too frequent and too much', 'regular but too little', 'pale and scanty but regular' and 'more frequent than natural'. The adjectives associated with blood loss ranged from 'scanty', 'profuse', 'excessive', 'small', 'irregular', 'disturbed', 'deficient' to 'slight', 'pale', 'small' and 'copious'. Notably, no records utilised positive terminology or present a story of healthy menstrual flow. More often than not, measuring menstruation was related to woman's sexual and social status as mother or potential mother. In this sense, it is not surprising that most of the asylum cases concerning menstruation relate to young, single women. As menstruation served as a primary indicator of fertility and nubility, the desire to control the menses can be read as a desire to guarantee one's status as a 'real' woman. This is reflected in the case reports in the *Lancet* in which the success of a cure for menstrual irregularity is marriage with details of the former patient's nuptials being reported to fellow colleagues as proof of the value and importance of gynaecology.

Rethinking menstrual paradigms

The medical profession should not, of course, be viewed as an uncontested or uniform body of opinion. From the 1870s onwards, research into the character of menstruation, its relationship to ovulation, menstrual metabolic rates and the effects of social and environmental factors on menstrual experience all facilitated alternative ways of thinking and talking about periods. In particular, debate among psychiatric and gynaecological specialists began to suggest that the term 'hysteria' was often misused and its relationship to the sexual organs simplified. Writing in the *Lancet* in 1881, James Totherick concluded that 'one might heartily wish that the word hysteria were banished from our medical language, and indeed, so far as its etymological signification is concerned, from our thoughts' (778–9). In particular, the term 'amenorrhoeal insanity' was increasingly deemed inappropriate, not least because it implied that all women with amenorrhoea were insane. As one practitioner noted in 1913, lunatics with heart afflictions were never said to have 'cardiac insanity'. In most cases, amenorrhoea was simply incidental to insanity (Macnaughton Jones *passim*). Increasingly, relationships between the body and the mind were described as too complex to be reduced to simplistic and potentially meaningless terms. Likewise, confidence in the routine conduct of gynaecological examinations in asylums wavered. In 1909 a female practitioner, Kate Hogg, examined the entry books for several asylums and concluded that a direct relationship between menstruation and insanity was rare (279). The proliferating uncertainty surrounding the relationship between menstruation, general well-being and mental health was inseparable from the movement towards campaigning for women's rights and political representation and the growing numbers of women entering the medical profession, not least because such women held a vested interest in proving that biology need not disable women's opportunities or rights to equality. Female practitioners were, perhaps, particularly well-placed to challenge male-authored paradigms of menstrual illness, not least on account of personal experience. In the 1870s, Elizabeth Garrett Anderson publicly rejected the idea that menstruation disabled women whilst, in America in 1887, Mary Putman Jacobi began to raise questions about the need for rest during the menstrual period. As female practitioners were keen to point out, they were much better qualified than male peers to write about menstruation on account of personal experience.

From the First World War, female-led clinical and social (qualitative) research into the character and understanding of menstruation flourished. Importantly, the new female-authored narratives of menstruation, especially those written by the Medical Women's Federation (MWF), aimed to re-create menstrual experience in a language of normality, activity, hygiene and ability rather than pathology, passivity and disability. Moreover, these narratives were created from understandings of menstrual experience as conceived and articulated by healthy women rather than from clinical contact with women defined as sick. For instance, Alice Sanderson Clow, a school medical inspector, initiated qualitative research into the attitudes amongst young girls towards puberty and menstruation. The inspiration for the project sprang from Clow's dissatisfaction with the terminology of menstruation, in particular the language of being 'unwell', unclean and cursed. For Clow, this language perpetuated self-fulfilling prophecies: girls took to their beds and missed lessons because, firstly, they expected to feel ghastly and, secondly, all too often the prospect of missing classes for a couple of days appealed to the adolescent. More importantly, however, such attitudes established a psychological framework which associated female biology with illness and laid down a pattern for a fertile lifetime of being metaphorically and literally unwell (511–13). Perhaps the most definitive expression of the approach adopted by practitioners like Clow was the publication of *The Hygiene of Menstruation: An Authoritative Statement* by the MWF in 1925, which explicitly rejected the paradigms of old based upon notions of menstrual debility as normative: 'Menstruation is a natural function; it is not an illness, and girls should therefore continue their ordinary work and play during the period. It should not be and is not normally accompanied by pain or malaise.'[22] The statement concerned itself largely with creating new frames of reference for menstruation (vitality, modernity, health, vigour, participation, energy) and was distributed to general practitioners and schools.

It would be naïve, however, to suppose that new models of menstruation simply replaced the old or that notions such as menstrual insanity were relegated to annals of superstition. Firstly, it could be argued that new narratives simply created new menstrual taboos. Notably, for those women who did articulate a menstrual experience associated with pain, tiredness or any form of distress, this discomfort now risked becoming taboo. Moreover, the new narratives of menstruation, disseminated through leaflets and booklets to general practitioners, schoolteachers, in magazines and inside sanitary products, were still tied to cultural notions of femininity. The deluge of information and the re-shaping of menstrual knowledge into a set of positive beliefs was intended to create a new menstrual experience where the menstruating woman could do 'almost anything': she could be clean, odour free, 'sweet and graceful' and dainty ('Menstruation'). Advertisements for Kotex, for instance, showed young girls dancing in the arms of handsome and strong men, seemingly as proof of emancipation from the problems associated with menstruation.[23] Yet this was not a version of femininity where doing 'almost anything' encouraged girls not to be train drivers or brain surgeons, but, rather, to aspire to a romantic ideal, to conceal menstruation and to be clean. New narratives simply rewrote a sophisticated version of menstrual etiquette, a version that was prescriptive in how to be a girl and how to 'cope' with periodic bleeding.

New menstrual narratives were also intended to provide a vocabulary for dialogues between mothers and daughters to counter ignorance and superstition. Yet this literature devalued the experience and teachings of mothers and grandmothers. Popular habits rooted in taking rest and/or avoiding water during the period were assigned to 'defective training'; euphemisms such as the curse, unclean and unwell were dismissed as ignorant (Clow 513; cf. Scottish Council for Health Education leaflet). An information sheet distributed inside the magazine *Woman* in the 1930s advised mothers not to enter into dialogue with their daughters concerning the period unless they were armed with new, rational knowledge. In particular, they were advised not to divulge their own experiences, especially if these were negative in any way.[24] Herein lies the tension in the new narratives of menstrual experience. Whilst encouraging a new culture of rationality, hygiene, health and communication, organisations such as the MWF risked overlooking the complexities and internal rationale of older customs and forms of menstrual discretion; they were promoting one vision of femininity and experience, the modern, at the expense of another. The second reason we should perhaps be suspicious of overestimating the impact of new hygienic and healthy models of menstruation is that the idea of women as peculiarly subject to their biology did not necessarily go away. Rather, it was spoken of in new ways. The work of Kyusaku Ogino and Hermann Knaus in the 1920s established that menstruation and ovulation were separate, rather than concurrent, events. The significance of this work for understandings of reproduction and the development of 'safe time of the month' contraception cannot be underestimated. Yet it also facilitated new paradigms of hormonal fluctuation. Directly related to this was the formulation of Pre-menstrual tension (PMT), an umbrella term for the physical, emotional and mental symptoms of menstruation, coined in 1931. How far this new paradigm emancipated women from the straitjacket of the Victorian model of menstruation remains hotly contested. As Sophie Laws has argued, reference to 'being hormonal' remains a common means of dismissing women in contemporary culture (22–3). That PMT remains a controversial issue indicates the ongoing ways in which gender is embodied.

Conclusion

Medical paradigms of menstrual experience in the nineteenth century explicitly rooted female biology within a domestic and procreative orbit whilst promoting notions of the inherent instability of female bodies and the vulnerability of female minds. Women could not expect to participate in society on an equal footing with men on account of putative scientific claims about the effects of 'feminine biology'. 'New' narratives of menstruation, created largely by women and in direct relation to campaigns for women's rights, utilised a vocabulary of modernity, health and hygiene in a bid to renegotiate menstrual and feminine identities away from the pathological paradigms of Victorian gynaecology. Yet it is possible to argue that far from freeing women from the strangulating menstrual napkin with which this chapter began, new narratives perpetuated a sophisticated but prescriptive paradigm of menstruation. Notably, early twentieth-century versions of menstrual etiquette remained tied, firstly, to conceptions of menstruation as problematic and taboo and, secondly, to perceptions of femininity that privileged marriage and motherhood. Finally, conceptions of emotional and physical periodicity, encapsulated in the concept of PMT, ensured the longevity of a 'scientific' narrative that cast women as the victims of their own bodies. Thus, whilst menstrual knowledge may have been in a state of flux at the turn of the twentieth century, what remained constant was the appropriation of the body as a field for the definition of sexual difference.

Notes

1. Gloucestershire Record Office (GlRO), Horton Road Asylum Box: HO22.
2. See Moscucci *passim*, and Hall *passim*.
3. See Tilt (February 1851); and Tyler-Smith (February 1856).
4. *Lancet* (29 October 1881): 772.
5. This could, of course, have been a euphemistic reference to a suspicion of pregnancy.
6. Gwent Record Office (hereafter GWO), D3202/2, 289–90.
7. GWO D3202/3, 251.
8. GWO D3202/1, 638.
9. GWO D3202/5, 127.
10. GWO D3202/5, 329.

11. GlRO HO22.
12. GlRO HO22.
13. GWO D3202/41/2.
14. GWO D3202/41/2.
15. GWO D3202/41/2.
16. GlRO HO22.
17. See Anon., 'Editorial' *passim*; Wade *passim*; Whitehouse, 1926 *passim* and 1927 *passim*.
18. GWO D3202/41/4.
19. GWO D3202/41/3.
20. GWO D3202/41/1.
21. GWO D3202/41/1.
22. Wellcome Library for the History and Understanding of Medicine. CMAC SA/MWF/H.51/2.
23. Wellcome Library for the History and Understanding of Medicine. CMAC SA/MWF/H.51/2.
24. Wellcome Library for the History and Understanding of Medicine. CMAC SA/MWF/H.51/2.

Works cited

Gloucestershire Record Office. Horton Road Asylum records.
Gwent Record Office. Pen-y-Fal Asylum records.
Wellcome Library for the History and Understanding of Medicine. Medical Women's Federation Files.
Anon. 'Editorial'. *Lancet* (26 September 1908): 950–1.
Anon. 'Menstruation and Pain'. *Lancet* (20 March 1926): 611–12.
Anon. Review of *Complete Ladies Guide to Physiology*. *Lancet* (5 August 1854): 103.
Anon. Review of 'On the Diseases of Women and of Ovarian Inflammation' by E.J. Tilt. *Lancet* (19 March 1853): 270–1.
Anstie, Francis, E. *Lancet* (14 December 1872): 839–42.
Barnes, Robert. *Lancet* (26 April 1867): 530.
——. *Lancet* (24 July 1880): 121–3.
——. *Lancet* (27 June 1896): 1797.
Clow, Alice Sanderson. *British Medical Journal* (2 October 1920): 511–13.
Hall, L. *Sex, Gender and Social Change in Britain Since 1880*. Basingstoke: Macmillan, 2000.
Hogg, Kate. *Lancet* (23 January 1909): 279.
Lansbury, C. 'Gynaecology, Pornography and the Anti-Vivisectionist Movement'. *Victorian Studies* 14 (1985): 413–37.
Laws, Sophie. 'Who Needs PMT?' *Seeing Red: The Politics of Pre-menstrual Tension*. Ed. Sophie Laws, Valerie Hey and Andrea Eagan. London: Hutchinson, 1985. 16–64.
Macnaughton Jones, H. *Lancet* (11 August 1900): 446.
——. *Lancet* (29 March 1913): 879–81.
Marion Sims, J. *Lancet* (4 March 1865): 224–5.
Martin, Emily. 'The Egg and the Sperm: How Science has Constructed a Romance Based on Stereotypical Male-Female Roles.' *Feminism and Science*. Ed. Evelyn Fox Keller and Helen Longino. Oxford: Oxford University Press, 1996. 103–14.

Maudsley, Henry. *Lancet* (9 May 1874): 663–4.

Medical Women's Federation. *The Hygiene of Menstruation: An Authoritative Statement.* Wellcome Library for the History and Understanding of Medicine. CMAC SA/MWF/H.51/2.

'Menstruation'. c. 1940. Educational Film. Wellcome Library for the History and Understanding of Medicine. CMAC SA/MWF/H.51/2.

Moscucci, Ornella. *The Science of Woman: Gynaecology and Gender in England, 1800–1929.* Cambridge: Cambridge University Press, 1990.

Poovey, Mary. *Uneven Developments: the Ideological Work of Gender in Mid-Victorian England.* London: Virago, 1988.

Routh, Amana. *Lancet* (27 June 1896): 1797.

Savage, George. *Lancet* (31 October 1903): 1209–13.

Scottish Council for Health Education. Leaflet. Wellcome Library for the History and Understanding of Medicine. CMAC SA/MWF/H.51/2.

Tanner, T. H. *Lancet* (15 October 1853): 363–4.

Tilt, Edward John. *Lancet* (8 February 1851): 148–9.

——. *Lancet* (22 November 1855): 564–6.

Totherick, James Y. *Lancet* (14 May 1881): 778–9.

Tyler-Smith, William. *Lancet* (2 February 1856): 113–15.

——. *Lancet* (4 June 1856): 113–14.

——. *Lancet* (4 June 1853): 509–10.

Usher Somes, L. N. *Lancet* (10 May 1890): 1010–11.

Wade, W. F. *Lancet* (5 June 1886): 1054–8.

Whitehouse, B. *Lancet* (1 August 1926): 382–4.

——. *Lancet* (18 June 1927): 1275–9.

9
Embryological and Agricultural Constructions of the Menstrual Cycle, 1890–1910

Helen Blackman

In 1915 the agricultural physiologist Francis Hugh Adam Marshall published an article entitled 'Physiology and Bacon-curing'. The publication was the culmination of several years work with colleagues at Cambridge University investigating the effects of ovariotomy (spaying) in pigs. Marshall asked two questions: did spaying make pigs put on the desired weight and did it affect the formation of black pigment in the abdomen, from where the best cuts of bacon are taken? Butchers claimed that sows killed during 'heat' had black pigment in the belly which made their meat difficult to cure and inferior in taste. Marshall and his co-workers had earlier concluded that spayed sows did put on weight faster than unspayed animals but that

> the black pigment so often found in the mammary area of sows belonging to coloured breeds is in no way related to sexual changes occurring during the period of heat or oestrus. On the other hand it is closely similar to, or identical with, the pigment of the hair, and is consequently harmless. (Marshall and MacKenzie, 'Ovariotomy' 413)

Marshall calculated that large bacon factories lost several thousand pounds a year because they would not use the belly cuts of meat from sows in heat and yet concluded that the pigment had little to do with the female cycle, and instead had everything to do with the colour of the pig's skin. Apparently spaying the animals when very young did prevent the formation of the pigment, but the pigment did not appear cyclically. Butchers, agriculturalists, and medical doctors drew on one another's beliefs in their assumption that all female physiology was dictated by a sexual cycle. Female animals, human and otherwise, were assumed to be governed by their reproductive systems even though little was known of the physiological details of reproduction until the early twentieth century. From the 1890s researchers from a variety of backgrounds, including agriculture, obstetrics, anthropology and embryology, began detailed investigations, and amongst this detail lay a great deal of speculation over the relationship between oestrus, menstruation, and ovulation.[1] To all this there was a decided asymmetry: the male reproductive system was rarely scrutinised nor was it assumed to dictate male behaviour.

This chapter examines the construction of histological – that is, cellular or microscopical – stages of the menstrual and oestrus cycles, and the homologies drawn between the two, using 'homology' to mean processes or organs which have the same evolutionary origin but which have perhaps developed different functions. While Emily Martin has charted the nineteenth-century view of menstruation in general as failed production (45–7), I will demonstrate that the construction of histological stages enabled, and was influenced by, complex and essentially ambiguous views of menstruation. Much depended on the interpretation of what happened during the menstrual cycle at a microscopical level, because this histological evidence was taken as the basis for verifying anatomical facts. The Cambridge embryologist Walter Heape labelled menstruation a degenerative process, but he was also astonished at the capacity of the uterus to recuperate its losses and likened menstruation to a preparation for oestrus. Heape was later to refer to spinsters as 'waste products of our Female population' who should not be allowed the vote (*Antagonism* 208). Marshall, although he was involved in the birth control movement, had little to say about women's position in society, and at first glance his science seems more neutral. Yet the stages he adopted for the menstrual cycle are just as much a construct as Heape's. My aim is to show that at a detailed, microscopical level, menstruation could be interpreted in many ways because political and medical beliefs interacted and were co-constructed. Changes in women's social roles becoming reflected in their bodies at the very deepest level.

Concerned about national supremacy, and picking up on the eugenic fears of contemporaries and co-workers such as Francis Galton, Heape became concerned more generally with reproductive issues. Biologists were beginning to train using the type system: since it was impossible to cover the entire animal and plant kingdoms in biological courses, one species came to represent the type of a number of other related species

(Geison 133–5). Thus, for Heape, Rhesus monkeys could be viewed as a type of primate, and information about their systems could provide information about other primates, including humans. As part of his research, Heape aimed to bring knowledge from animal breeders into the scientific realm, and he began by clarifying the terminology they used. He particularly disliked the use of the term 'heat' and argued for a distinction between oestrus, or the period of sexual desire, and pro-oestrus, or the histological changes which enabled oestrus. It was this distinction that enabled Marshall's interpretation of the menstrual cycle as a process of regeneration, yet, as I shall argue, and as Heape admitted, his periodisation 'cannot be quite sharply defined' and one stage often blended into the next ('Entellus' 462).[2] Marshall drew on Heape's work, using the stages worked out for primates as a template for understanding the reproductive cycles of a number of species including the dog, ferret, and sheep. Yet Marshall pointed out that not all species comfortably fitted Heape's model, inadvertently demonstrating the number of possible constructions of reproductive cycles. Although Marshall used the comparisons between the stages in different species to confirm the pro-oestrus homology, the stages of the female cycle were clearly imposed, rather than discovered. The Scottish biologists Patrick Geddes and J. Arthur Thomson argued that ancient theories of menstruation could be divided into two main categories – the theory that menstruation was a freshening of the womb and the idea that it was a mini-abortion (245). Marshall briefly revivified the freshening thesis for the twentieth century, yet his idea was short lived. As histological work gave way to the study of hormones, or endocrinology, menstruation came to be viewed once more as a failure, and no longer as homologous to a preparatory stage.

Heape and the construction of the menstrual cycle in monkeys

Since the ancient Greeks, Western doctors had understood the female body as built around the uterus. During the nineteenth century, attention switched to the ovaries as investigators speculated on the timing of ovulation in humans and other animals. In the 1840s the zoologist Felix A. Pouchet argued that ovulation occurred during oestrus and that menstruation was analogous to oestrus, thus giving the ovaries a new and vital role in menstruation and inviting a series of animal/human comparisons. Since menstruation was the most obvious part of women's cycle, and oestrus or heat was the most conspicuous part of the cycle in other animals, medical doctors and biologists commonly believed that the two were closely related. The extent to which attitudes to women were interwoven with attitudes to other female animals became more apparent after Pouchet's work, for just as a sow's meat was believed to be off-colour during heat, so it was commonly believed that the touch of a menstruating woman could turn meat sour. Indeed many behavioural changes were associated with the cyclical nature of menstruation, in both popular and scientific discourses. As one renowned sexologist put it, 'while a man may be said, at all events relatively, to live on a plane, a woman always lives on the upward or downward slope of a curve' (Ellis 248).

Despite his contemporaries' increasing conviction that it was the ovaries that held the key to understanding the female reproductive cycle, Heape began by investigating the histology of the uterus. Contemporaries were also starting to believe that the body was governed to an as-yet-unexplored degree by secretions from the ductless glands. These secretions, named 'hormones' in 1905, were believed to send messages through the blood to remote parts of the body, offering an endocrinological explanation for the reproductive cycle rather than the mechanical and neurological explanations common for most of the nineteenth century (Borell 267–8; Oudshoorn 16; Sengoopta 425). Concerns about degeneration, race hygiene and international competition brought more focus on motherhood and reproduction, placing increased pressure on gynaecologists and biologists to map the physiology of the uterus. By the end of the First World War, researchers described the menstrual cycle as a specialised example of the oestrus cycle, and both were thought 'governed' by hormones secreted by the ovaries and *corpora lutea* (Corner, 'Knowledge' 919). It took substantially more time for this information to appear in textbooks. Although much of this work took place on mainland Europe, it was Heape who fully established the terms 'oestrus' and 'pro-oestrus' and who asserted a scientific definition of reproductive terminology over that commonly used by animal breeders.

From the late 1870s Heape worked at Cambridge University with the renowned embryologist Francis Maitland Balfour. Like many biologists of his time, Heape received little formal training, but his work is typical of the Cambridge school of zoology. The school initially concentrated on what later critics called 'descriptive' embryology, detailing the development of different species but not actively interfering in the growth of the embryo. Balfour used the latest embryological techniques in preservation and preparation to stage the development of the chick and the elasmobranch fishes, promoting this as the correct methodology for comparative embryological work. Ageing an embryo, when the exact time of fertilisation was unknown, proved a complex problem. Individual embryos of the same species develop at different rates, and organs which appear early in one may not appear at precisely the same time in another. Nor is size a reliable guide to age. So Balfour devised a method for comparing embryos at different stages. Heape's staging of the menstrual cycle echoes the work of his mentor and, just as embryonic stages were an uncertain construct, so were the stages of the female cycle. After initial work on the embryology of the mole, Heape turned his attention to wider problems with reproduction and began to investigate the possibility of embryo transfer and artificial fertilisation (Biggers 174–7). Much of Heape's work had economic and agricultural implications, as is evident from his correspondence with Galton, who proposed that Heape should begin experiments with artificial fertilisation. Galton was concerned that while 'the number of them secreted by the male is enormous', as it only took one spermatozoon to fertilise an ovum, 'a vast quantity of material is wasted'. This meant that a breeder could not take full advantage of any sport (genetic freak) produced. Artificial fertilisation would allow a minute proportion of spermatozoa to produce a large number of offspring, and thus 'a new and valuable power would be placed under the control of breeders'. Heape agreed, but in pursuing this work he came to realise that even the basics of reproductive physiology – such as the histology of the uterus – were largely unknown.

During the 1880s the obstetrician John Williams had attempted to plot the histological changes in the uterus, but he could find very little human material. Williams could not use unhealthy uteri removed from otherwise healthy women in deciding upon physiological changes, so he removed uteri from dead women. His critics seized upon this fact, arguing that his results were inaccurate for he relied on material they deemed pathological – indeed it proved impossible to decide if a mortally ill woman could possess a healthy uterus (Blackman ch. 2). So in deciding to study menstruation, Heape circumvented this problem by researching on monkeys, which he saw as acceptable models for humans. As he commented in his December 1889 application for funds,

> I have been reading the literature of this latter subject for the last few months and am much impressed with the chaotic condition of the minds of medical men & with the immense importance of the subject, and I feel very sanguine that a good understanding of the menstrual phenomena in monkeys will be of great use to medecine [sic]. (Letter to Professor Newton)

Heape succeeded in gaining £100 to ship forty Macacus Rhesus from India to England. The monkeys were, however, too young for breeding experiments, and Heape was informed that older monkeys were too vicious to ship together, but would need expensive separate crates (Heape, 'Entellus' 412). It would be cheaper for him to go to India. In September 1890 Heape applied for a Balfour Studentship, tenable for three years. As he explained to the selection board in September 1890,

> Should I be elected to the Studentship I would propose to continue on a more extensive scale my experiments upon the transplantation of ova, and (as in all experiments upon breeding) there would be pregnant periods of inactivity I should propose to work also at the Embryology & sexual phenomena (e.g. menstruation) of monkeys. (Letter to Selection Board)

With the world's first successful embryo transplant behind him, Heape was awarded the studentship. Yet he encountered two further problems in India. Illness, and the fact that he could not collect monkeys in areas where they were sacred, frustrated him and, advised by a doctor to return home, he culled the monkeys in India (Heape, 'Entellus' 414). He brought back preserved uteri, deciding to abandon his work on embryos, since too few of the monkeys were pregnant, and to concentrate instead on their menstrual phenomena. Since the monkeys were represented by their internal reproductive organs alone, any histological changes were not related to their general physiology or to their behaviour.

Heape sorted the uteri into four periods, further subdivided into eight stages, as follows:

A. Period of rest
Stage I. – The resting stage
B. Period of growth
Stage II. – The growth of the stroma
Stage III. – The increase of vessels
C. Period of degeneration
Stage IV. – The breaking down of vessels
Stage V. – The formation of lacunae
Stage VI. – The rupture of lacunae
Stage VII. – The formation of the menstrual clot
D. Period of recuperation
Stage VIII. – The recuperation stage ('Entellus' 415–16)

These stages were decided upon according to the histology of the uteri, with Heape ensuring that he had at least three or four specimens for each stage. Although there had been piecemeal work on other species, Heape was the first to carry out such a thorough study on primates, and, although some of the details were refined, his work was often cited approvingly in the succeeding decades. He was quite happy to allow the monkeys to be used as substitutes for humans, attributing any differences between his findings and those of gynaecologists to the facts that his material was drawn from healthy uteri and that there was more of it, rather than to differences between the species (Heape, 'Entellus' 450). This was not an unusual methodology as it was widely accepted that organ systems bore strong resemblances across different species. Thus Heape was investigating the phenomena of menstruation, and the fact that he chose monkeys was only a secondary consideration.

As with his embryological work, these stages were an abstraction, an attempt to define 'a normality'. Heape declared that he could not assert that the stages 'invariably occur in the same individual, or in another individual to the same extent' ('Entellus' 450). Neither were they clearly marked when they were seen, for the recuperative stage actually began during the previous stage, the formation of the menstrual clot ('Entellus' 462). Although the construction, and labelling, of what counts as 'normal' was, and is, common throughout the biological sciences, here we can see that much depended on the interpretation of the female body: those who advocated equal education for women and men were particularly anxious to prove that activities beyond the reproductive sphere did not have adverse effects on female physiology. Heape, though, was arguing that not to reproduce would result in degeneration.

Since, prior to Heape's work, menstruation was likened to oestrus, doctors commonly believed that menstruation and ovulation were coeval, and even that menstruation might bring about ovulation, since pressure from swollen tissue may cause the Graafian follicle containing the egg to burst. Unsure about the relationship between the two, these doctors did not simply believe that menstruation was a sign of failed ovulation. Heape, examining monkeys' ovaries in relation to their uteri, concluded that menstruation and ovulation were not related. He speculated that increased pressure during menstruation might bring about ovulation, and that it was possible that the two had the same cause, but argued that doctors were mistaken in their belief that menstruation and ovulation were concurrent. Heape then developed these ideas, examining other species and relating histological changes in the uterus to menstruation and ovulation. In his next batch of publications on menstruation, in 1897 and 1898, he argued that histological evidence pointed to an analogy between heat and the early stages of menstruation ('Human Female' 167). He refined these ideas further when he went on to define the terminology used to describe various aspects of reproduction. Heape believed that this terminology was confused because people used it in so many different ways, so he set out to differentiate between a breeding season and a nursing season, and argued that the term 'rut', used by some breeders to denote female heat, should only apply to male animals which exhibit seasonal sexuality, not to females or to males who breed at all times of the year. He was particularly concerned to emphasise the difference between oestrus and pro-oestrus, and his terms are still used today. Breeders used 'heat' and 'oestrus' interchangeably to denote a series of changes in the cycle which, he argued, should be separated. Oestrus proper referred only to the time when the female was receptive to the male. The pro-oestrus consisted of preparation for this time, during which the uterine mucosa underwent the changes necessary for oestrus. The pro-oestrus was characterised by bleeding in some species, most notably the bitch and the cow, and it was thus the pro-oestrus alone, not oestrus, which was synonymous with menstruation (Heape, 'Sexual Season' 45).

Marshall and the pro-oestrus homology

Marshall acknowledged his debt to Heape, who helped him in the early stages of his career. Both men touted the practical implications of their work, but whereas Heape presented his research to the Royal Society and the Obstetrical Society, Marshall published in a variety of agricultural and biological journals. He used Heape's work as a template, investigating the reproductive cycle in ferrets, dogs and sheep. As with his later work on pigs, this was intended to have very practical applications for all species. Marshall sat the Natural Sciences Tripos in Cambridge in the late 1890s, and his work bears strong resemblances to Heape's, particularly when he used types of animals and made comparisons between species. Such comparisons were a major part of the Cambridge school's project, as they looked for evolutionary links. Hence whilst Heape used entellus and rhesus monkeys to represent all primates, Marshall used sheep to represent all ovines and domestic dogs to represent all canines. In the early 1900s, Marshall concentrated on working on sheep, relating behaviour during the oestrus cycle and the histology of the uterus to Heape's eight stages and to 'similar changes in other types' (Marshall, 'Sheep' 56). Working in the zoological and physiological departments of

Edinburgh University, Marshall examined *corpora lutea* – the yellow bodies which form when a Graafian follicle ruptures to produce an egg, now known to secrete hormones which influence pregnancy and the menstrual cycle. In 1905, he postulated that oestrus and pro-oestrus were brought about by an ovarian secretion, and that the *corpus luteum* also secreted something ('The Oestrus Cycle' 340). Marshall was trying to fit the uterine cycle of all species into one model, whether or not that species had a menstrual cycle or an oestrus cycle, which at that point were not clearly differentiated. Indeed, the gynaecologist and zoologist George W. Corner claimed, 'Marshall had the great advantage of knowing the female cycle by observation of domestic animals, the ewe, ferret and bitch, rather than of women, whose cycles, marked by menstruation instead of oestrus, were still imperfectly understood by the puzzled clinicians' ('Early History' vii).

Although Marshall referred in his later work to the pro-oestrus period in sheep as an established fact, it is clear from his research papers that it had not always been so. Marshall stated, 'so rapidly does the œstrus follow the prooestrum, that in some cases where the [histological] changes characterising the prooestrum are slighter than usual, the two stages of the cycle seem almost to become one stage' ('Sheep' 48). External signs were no more reliable, for the most marked of these, the mucous flow, often continued during the oestrus and metoestrus periods ('Sheep' 46). The oestrus period proved similarly elusive. Its most obvious external sign, the animal's behaviour, was not always coincident with its most obvious internal sign, ovulation. As Marshall knew, bats appeared to go through oestrus and mate several months before they actually ovulated and conceived ('Sheep' 70–1). There was no one model of the oestrus cycle that could fit all animals.The confusion became even more marked the more species Marshall included in his research. When studying the ferret he stated that, 'the œstrus associated with the swelling of the vulva extends far into the uterine recuperation stage, if not sometimes beyond it; so that similar changes to those which in the sheep characterise metoestrum, in the ferret may occur contemporaneously with œstrus' ('Sheep' 64). So oestrus for the ferret may be metoestrum (the period of 'rest') for the sheep. Different species clearly had differing cycles, but they were more easily understood if the reproductive cycle was held to unify all female animals. Marshall admitted these differences when he remarked that

> all the stages into which Mr Heape first divided the menstrual cycle of *Semnopithecus* are represented in the œstrus cycle of the bitch ... , and

in that of the ferret, with the exception of Stage VII. (the formation of the menstrual clot), the characters of which in the case of the latter animals are only partially recognisable. It is apparent also that the prooestrus stage in the carnivore, in regard to its severity, is approximately intermediate between that of the sheep, as described above, and the menstruation of monkeys. ('Sheep' 65)

The stages, not even quite clear to Heape in his work on primates, could only be applied to other species if considerable flexibility was allowed. Yet a great deal depended on the comparison of the menstrual and oestrual cycles: if menstruation could be likened to the pro-oestrus, then menstruation was a preparation, but when exactly did the pro-oestrus occur? As Marshall admitted, 'the matter is complicated by the undoubted fact that in one part of the uterus a more advanced stage may be reached than in another part of the same uterus' ('Sheep' 60). The periods did not map clearly onto one another and, even within the same uterus, different stages might be present at the same time. Clear stages did not occur 'naturally', but were constructed by Marshall and Heape in their attempts to understand mammalian reproductive cycles.

In the first edition of *The Physiology of Reproduction* Marshall stated that menstruation and pro-oestrus were homologous and, for a short time, the homology became orthodoxy. Based on his own research, and his reading of others', Marshall argued that the bleeding stages, pro-oestrum and menstruation, were physiologically identical and histologically similar in the menstrual and oestrus cycles. The bleeding stage in non-human animals was held to be preparatory to ovulation, and so, by extension, menstruation became preparatory. Marshall was a little cautious about this theory, indicating that it was not necessarily true but that it was 'not opposed to any of the known facts' (*Physiology* 161–3). Despite this caution, the homology did enable Marshall to give an essentially positive interpretation of menstruation in women; the 'freshening' theory to which Geddes and Thomson referred was reborn for the twentieth century. However, by 1927 Marshall had changed his mind about the true nature of the pro-oestrus, and the homology had to change. Heape had labelled the period after oestrus in the bitch and other animals with a similar cycle as the metoestrum, or 'period of rest'. Such animals were described as mono-oestrus, that is, once they had experienced oestrus, if they did not become pregnant they went through a period of rest. Animals such as the horse were polyoestrus, for if they did not become pregnant after oestrus, they had only a very short period of rest before they experienced oestrus again (Heape, 'Sexual Season' 9).

Yet once the corpus luteum's role in pregnancy was more fully researched, the period immediately following oestrus in 'non-pregnant' animals was labelled 'pseudo-pregnancy'. Oestrus in the bitch, Marshall stated, was followed by 'a period of development of the corpus luteum, associated with growth of the uterus and mammary glands' ('Pseudo-pregnancy' 205). Since similar changes occurred in pregnancy, the pro-oestrus came to be labelled 'pseudo-pregnancy'.

By this time it was known that ovulation in humans took place about the fourteenth day after menstruation began. Drawing on the work of other researchers, Marshall stated that the changes in the uterus at about that time were like those which occurred in pseudo-pregnancy in some lower mammals ('Pseudo-pregnancy' 207–8). Yet Marshall felt that histological evidence and the development of the corpus luteum meant that the pro-oestrus process in the bitch was 'something less than the complete menstrual phenomena of man' (208). Marshall also knew that in monkeys – and, he suspected, in humans – menstruation could occur in the absence of ovulation, and therefore in the absence of a corpus luteum. Reluctant to entirely abandon the homology between the pro-oestrus and menstruation, Marshall argued that menstruation was probably equivalent to both the preparatory period of pseudo-pregnancy and the degenerative changes which occurred during the pro-oestrus. His pupil Alan Sterling Parkes was less sanguine, always insisting that menstruation was equivalent to the end of pseudo-pregnancy (Parkes 244). The possible interpretation of menstruation as a preparatory process did not last long once physiologists began to draw comparisons between pro-oestrus and pregnancy, rendering the former a false attempt at the latter.

Conclusion

Heape was one of the first researchers to describe clearly the histological changes which take place in a primate uterus. He divided the menstrual cycle into stages, which he argued were marked and real, even though more than one stage could be present in one uterus. By labelling the menstrual stage degenerative Heape fed into, and drew upon, an essentially negative view of menstruation. However, his stages allowed Marshall, albeit briefly, to resurrect a freshening theory of menstruation, rather than portraying it as a mini-abortion. But scientists believed that the purpose of the female cycle could only be pregnancy, and menstruation came to be likened to a pseudo-pregnancy, showing the persistence of negative portrayals of menstruation. I am not arguing that scientists have not grasped the reality of menstruation. Rather, I have shown that menstruation can be construed in a number of ways. Even in complex, technical work, perhaps especially in complex, technical work, there is room for such interpretation. The oestrus and menstrual cycles were believed to link all female animals, of whatever species, and the work of Heape and Marshall shows the problems encountered when scientists do make comparisons across species – menstruation could be degenerative, or preparatory or a pseudo-pregnancy, depending in part on the view of the observer.

Notes

1. The meaning of oestrus was not yet fixed and could be applied generally to the period of 'heat' or more specifically to the time when an animal ovulated.
2. My thanks to UCL Library Special Collections and the Syndics of Cambridge University Library for permission to use material held in their collections.

Works cited

Biggers, John. D. 'Walter Heape, FRS: A Pioneer in Reproductive Biology. Centenary of his Embryo Transfer Experiments'. *Journal of Reproductive Fertility* 93 (1991): 173–86.

Blackman, Helen. 'Women, Savages and Other Animals: The Comparative Physiology of Reproduction 1850–1914'. PhD Diss. Manchester University, 2001.

Borell, Merriley. 'Organotherapy, British Physiology and the Discovery of the Internal Secretions'. *Journal of the History of Biology* 9 (1976): 235–68.

Corner, George W. 'Our Knowledge of the Menstrual Cycle, 1910–50'. *Lancet* (28 April 1951): 919–53.

——. 'The Early History of Oestrogenic Hormones'. *Journal of Endocrinology* 31 (1964–5): iii–xvii.

Ellis, Henry Havelock. *Man and Woman*. London: Walter Scott, 1894.

Galton, Francis. Draft of a letter to W. Heape, 28 June 1894. Galton Papers, University College London. No. 140.

Geddes, Patrick and John Arthur Thomson. *The Evolution of Sex*. London: Walter Scott, 1889.

Geison, Gerald. *Michael Foster and the Cambridge School of Physiology: The Scientific Enterprise in Late Victorian Society*. Princeton: Princeton UP, 1978.

Heape, Walter. Letter to Professor Newton, 6 December 1889. Balfour Memorial Fund, Minute Book of the Managers, Cambridge University Library, Dept. of Manuscripts and University Archives, Zoo 10/1.

——. Letter to the Selection Board for the Balfour Studentship, 26 September 1890. Balfour Memorial Fund, Minute Book of the Managers, Cambridge University Library, Dept. of Manuscripts and University Archives, Zoo 10/1.

——. 'The Menstruation and Ovulation of Macasus Rhesus, with Observations on the Changes Undergone by the Discharged Follicle, Part II'. *Philosophical Transactions* B 188 (1897): 135–65.

——. 'The Menstruation and Ovulation of Monkeys and the Human Female'. *Transactions of the Obstetrical Society of London* 40 (1898): 161–74.

——. 'The Menstruation of Semnopithecus Entellus'. *Philosophical Transactions* B 185 pt. I (1894): 411–67.

——. *Sex Antagonism*. London: Constable & Co., 1913.

——. 'The "Sexual Season" of Mammals and the Relation of the "Pro-oestrum" to Menstruation'. *Quarterly Journal of Microscopical Science* 44 (1901): 1–69.

Marshall, Francis Hugh Adam. 'The Oestrus Cycle and the Formation of the Corpus Luteum in the Sheep'. *Philosophical Transactions* B 196 (1904): 47–97.

——. 'The Oestrus Cycle in the Common Ferret'. *Quarterly Journal of Microscopical Science* 48 (1905): 323–45.

——. 'On the Pro-oestrum and Pseudo-pregnancy'. *Quarterly Journal of Experimental Physiology* 17 (1927): 205–10.

——. *The Physiology of Reproduction*. London: Longman, 1910.

Marshall, Francis Hugh Adam. and K.J.J. MacKenzie. 'Physiology and Bacon-curing'. *Journal of the Royal Agricultural Society of England* 76 (1915): 1–13.

——. 'On Ovariotomy in Sows with Observations on the Mammary Glands and the Internal Genital Organs, part 1'. *Journal of Agricultural Science* 4 (1912): 410–20.

Martin, Emily. *The Woman in the Body: A Cultural Analysis of Reproduction*. Milton Keynes: Open University Press, 1989.

Oudshoorn, Nelly. *Beyond the Natural Body: An Archaeology of Sex Hormones*. London: Routledge, 1994.

Parkes, Alan Sterling. 'Francis Hugh Adam Marshall, 1878–1949'. *Obituary Notices of Fellows of the Royal Society* 7 (1950): 239–51.

Sengoopta, Chandak. 'The Modern Ovary: Constructions, Meanings, Uses'. *History of Science* 38 (2000): 425–88.

10
Of Sex, Nationalities and Populations: The Construction of Menstruation as a Patho-Physiology

Zahra Meghani

Williams Obstetrics: R.V. Short and Katharina Dalton

Endocrinopathy

In North America, *Williams Obstetrics* has the status of an epistemic authority on women's reproductive physiology. Since the 1980s, every edition of this obstetrics textbook has portrayed monthly menstruation as an evolutionary abnormality. In the nineteenth edition in 1993, the editors claim that because monthly menstruation is not the evolutionary norm for non-human female primates *and* that because some women (percentage unknown) experience the socially debilitating PMS (*Williams* 97), women's periodicity is an evolutionary aberration that is pathological.[1] They characterise it as an endocrinopathy, that is, an inherent hormonal disorder:

> With all good intentions, we have reassured our daughters that cyclical menstruation is normal; and from this perspective, it has been assumed by many women that repetitive menstruation is equated with femininity. If this prevalent premise that recurrent ovulation is the biological evolutionary norm were incorrect, it would be easier to accept the likelihood that the recurrent secretion and withdrawal of progesterone is an endocrinopathy that is pivotal in the development of luteal phase disabilities (i.e. PMS). (*Williams* 98)

Their implication is that menstrual suppressant contraceptives ought to be used as 'treatment' for all females of childbearing age, barring those contemplating pregnancy.

The editors of *Williams Obstetrics* owe this theory to R.V. Short, a British veterinarian. Short's 1976 'The Evolution of Human Reproduction' argued that in evolutionary terms monthly menstruation was a recent phenomenon attributable to the prehistoric agricultural revolution. The increased availability of food heralded a rise in human female fertility, followed by the development of various means of contraception. That development, in conjunction with better diet, resulted in human females experiencing regular monthly menstrual cycles. Short believed that monthly menstruation in human females must be considered a pathological evolutionary aberration because:

- humans are primates and non-human female primates do not experience monthly menstruation;
- women in 'primitive' societies do not menstruate monthly;
- menstruation is correlated with the incidence of certain forms of cancer;
- menstruation means blood loss which could seriously compromise malnourished women in developing nations;
- women with PMS experience impairment in their ability to function well socially.

Short believed that his analysis of the 'problem' of monthly menstruation 'highlighted the importance of developing a non-steroidal contraceptive that would allow a woman to return to the reproductive state that was the norm for our primitive ancestors – amenorrhoea' (Short, 'The Evolution' 20). Of the various reasons that Short gave for his position, his PMS argument is the least speculative (Short, 'The Evolution' 19). Short's central claims about PMS were derived from the work of a British psychologist and general practitioner, Katharina Dalton. Dalton, researching in 1960, noted that out of a group of 217 11–17 year old girls, the performance of 54 girls in academic and housekeeping tasks suffered during their pre-menstrual phase and then rose to its former level in the week following menstruation. Dalton attributed the change in academic performance to water retention during the pre-menstruum. She concluded that, '[a]bout one girl in every six in any examination entry will be in her pre-menstruum and thus at her lowest intellectual ebb' ('Effect' 328).

Dalton's 1961 'Menstruation and Crime' argued that there was a significant correlation between the pre-menstruum and criminal behaviour. Her claim was based on a six-month-long study of a group of 156 regularly menstruating women prisoners. Dalton claimed that those 94 prisoners reported for misbehaviour were either in the pre-menstruum period or menstruating. Citing her earlier work, she noted that there was a marked similarity between 'the effect of menstruation on naughty schoolgirls, newly convicted women, and disorderly prisoners' ('Menstruation' 1753). She speculated that menstrual hormonal changes probably made these females 'less amenable to discipline' and thus inclined to misbehaviour or criminal activity ('Menstruation' 1753). But she also acknowledged the possibility that the symptoms of PMS, including 'lethargy, slower reaction time and mental dullness', may have made these females' anti-social activities more easily detectable than otherwise, accounting for increased reports of miscreant behaviour ('Menstruation' 1753). The problems with Dalton's research are numerous, ranging from a lack of control groups to the use of non-representative samples to unjustified conclusions to unwarranted universalisations.[2] It can be argued that the cascade of conclusions, from Short's larger conclusion that monthly menstruation was an evolutionary aberration to the *Williams Obstetrics'* categorisation of menstruation as an endocrinopathy, draws, therefore, on at least one justification that is seriously lacking in credibility.

A feminist analysis also reveals that these theorists' recourse to ideas of evolutionary suitedness is strategic. They do not recommend that men's reproductive physiology be aligned with the evolutionary norm for non-human primates, nor do they expect aspects of human physiology common to men and women to conform to the non-human primate standard. Bi-pedalism, for instance, is not categorised as a pathological evolutionary aberration although it is both unique to humans *and* causes some percentage of the population to experience various physical ailments, such as fallen arches and back pain, impairing their ability to function well in society. However, Short, Dalton and the editors of *Williams Obstetrics* have been quite willing to consider non-human female primate reproductive physiology as the norm to which women ought to aspire. This attitude betrays a commitment to a worldview that construes women as physiologically deficient. Their rhetoric is all the more troubling because they identify themselves as participating in a campaign for female health. Furthermore, by arguing for the use of menstrual suppressant contraceptives as 'treatment' for monthly menstruation, Short and the editors of *Williams Obstetrics* are able to align themselves with those who support reproductive rights, thus construing any critique of their position as motivated by anti-choice and anti-birth-control sentiments that compromise women's well-being.

Contextualising Short's theory

Short's endorsement of menstrual suppressant contraceptives as 'treatment' for menstruation takes on an interesting hue when viewed in light of his concerns about population growth. Four years prior to his menstruation paper, Short, deeply influenced by Robert Malthus's hypothesis that unchecked population growth would outstrip food supply, resulting in widespread misery, had published 'Reproduction and Human Society', arguing that the rapid growth of population in Asia and Africa posed a serious threat to the survival of the human species. Short maintained that the current accumulation of wealth by Western nations at the cost of nations in Asia and Africa, coupled with the rapid growth of population in Asia and Africa, had set the stage for global conflict with potentially disastrous results.[3]

> Many economists have predicted that the rich nations of the world are destined to grow richer, at the expense of the poor nations who are doomed to become poorer. It seems doubtful if mankind could survive the ensuing racial and national tensions if there was to be an increasing polarisation between the haves and the have-nots. ('Reproduction' 124)

Short believed that this global crisis could be averted if the West developed population control strategies, specifically contraceptives, for the female populations of Asia and Africa. In 1972, Short was not alone in expressing this populationist anxiety, inflected by racialising tendencies, about the growth rate of population in Asia and Africa; he was vocalising sentiments that had been echoing across Britain for more than a decade.[4] Beginning in the 1950s, an acute labour shortage in Britain had occasioned an in-flux of semi-skilled workers from former British colonies in South Asia and Africa. Most of these immigrant labour populations were concentrated in those industrial areas experiencing labour shortage.

Elizabeth Overton argues that the high fertility amongst these localised commonwealth immigrant populations, in contrast to the low and declining fertility amongst earlier white immigrant groups, coupled with stark differences in social practices and the resolution of the labour shortage, accounted for the public outcry amongst the white working class citizens opposing the presence of non-white new immigrant:

> Where jobs and housing were already in short supply, the employed 'coloured' population was accused of taking jobs which 'should have been given' to the indigenous populations, and keen resentment was

felt, for example, against immigrants with large families who thereby earned more points on council house waiting lists and were accused of 'jumping the queue'. (55–6)

Although the 'jumping the queue' claims were not justified, and although the newer immigrants were more economically vulnerable in periods of high employment than the white citizenry, the British government, in response to these nationalist tirades, implemented the first in a series of four immigration acts aimed at closing off Britain to South Asian and African immigrants (Overton 56). With the passage of the 1962 Commonwealth Immigration Act, the British government began both collecting data on the flow of people from its former colonies into Britain and limiting entry to those commonwealth citizens who held work vouchers, their dependents and the dependents of those already settled in the UK. Until the passage of this Act, citizens of the British Commonwealth nations had not been regarded as aliens, and, therefore, they had not been subject to immigration control by the British government. The Act was extended in 1968 to include East-African Asians and two more immigration control acts were passed in 1969 and 1971.

Short's 1972 paper on the global population crisis cannot, therefore, be read as having no local significance. In fact, his 1976 'Evolution of Human Reproduction', prescribing more or less universal use of menstrual suppressant contraceptives, could be read as an attempt to solve the 'problem' of high rates of reproduction amongst non-white populations within Britain and globally. The use of contraceptives for all women, barring those contemplating pregnancy, certainly appears politically motivated when transposed against this image of spontaneous non-white reproduction. The populationist concerns that appear to have motivated Short's theory are also present in *Williams Obstetrics*. Much like Short, the editors of this textbook construe 'over-population' to be 'the greatest hazard to the health and environmental and economic future of humankind' (*Williams* 11).[5] Whilst they do not identify Asia and Africa by name as regions with high population growth, they note with alarm that '[t]he population doubling time in some countries today is believed to be less than 20 years, perhaps as low as 12 or 15. The population of the world has already passed the 5 billion mark!' (*Williams* 11). However, disturbing as *Williams Obstetrics'* endorsement of universal use of menstrual suppressant contraceptives is, it is somewhat reassuring that it seems to not have received much uptake amongst the US medical community. Not all gynaecological and obstetrics textbooks discuss the similarity between women's reproductive physiology and that of non-human primates, and those that do, such as *Kistner's Gynecology & Women's Health*, do not go on to posit the reproductive biology of non-human primates as the norm that human females ought to conform to (25). However, in at least some US medical schools, future generations of gynaecologists and obstetricians are cutting their teeth on *Williams Obstetrics*, learning to view women through the lens of a theoretical paradigm that casts them as intrinsically flawed.

Is Menstruation Obsolete?: Sheldon Segal and the Population Council

Iron-deficiency anaemia and PMS

Distinct from the attempt in *Williams Obstetrics* to characterise monthly menstruation as a patho-physiology, thus necessitating the uniform use of menstrual suppressant contraceptives, another movement construing women's periodicity as a health risk has emerged in the last few years. Its proponents are Elsimar Coutinho, a Brazilian gynaecologist, and Sheldon Segal, an American biochemist and embryologist. In *Is Menstruation Obsolete?* (1999), they claim that unless contemplating pregnancy all females of childbearing age ought to use menstrual suppressant contraceptives. Strongest and foremost amongst their many reasons for recommending 'treating' monthly menstruation with menstrual suppressant contraceptives are the two arguments that it is responsible for PMS and that it both causes and compounds iron-deficiency anaemia. Coutinho and Segal claim that iron-deficiency anaemia is a serious problem for developing nations, and that the female population in these parts of the world is particularly compromised by anaemia because of menstrual blood-loss (92). As a solution to this problem they propose the use of menstrual suppressant contraceptives by all females of childbearing age:

> The problem of anemia in less affluent countries of eastern and southern Europe and in many developing countries remains much worse [than that in the richer, more developed nations]. Considering that the monthly loss of blood is absolutely needless and that the treatment of anemia significantly increases the cost of the already overburdened health services of these countries, the suppression of menstruation would be an important public health measure, as well as a health benefit to individual women. (160–1)

There are numerous problems with this proposal. The predominant cause of iron-deficiency anaemia amongst the poor of the Third World is inadequate sanitation facilities, resulting in high incidence of hookworm disease ('Parasitic' par. 23–5).[6] Presumably Coutinho and Segal would not be opposed to treating the root cause of iron-deficiency anaemia by providing the compromised populations with adequate sanitation facilities, iron tablets and treatment for hookworm disease, the latter solution, according to UNICEF, costing only a few US dollars per year per person ('Progress' par. 7). It is unclear how suppressing monthly menstruation is any part of a sound solution and quite clear how awesome a potential market the Third World can be for menstrual suppressant contraceptives.

Coutinho and Segal also argue that menstrual suppression should be adopted by women in wealthy, industrial, nations such as the US, because, according to a study published in the *Journal of the American Medical Association*, about 7.5 million young females in the US are iron-deficient (Coutinho and Segal 160).[7] They take this as evidence of a direct causal connection between monthly menstruation and iron-deficiency. However, their attempt to marshal that study in support for their position is problematic. The authors of the research project in question came to a conclusion that was inconsistent with Coutinho and Segal's thesis. In 'Prevalence of Iron Deficiency in the United States', Anne Looker *et al.* argue that 'iron deficiency and iron deficiency anaemia are still relatively common in the Unites States among women of childbearing age, especially those who are black or Mexican American, poor, and have 12 or fewer years of education or 4 or more children' (973). Coutinho and Segal's claim that there is a simple, uni-directional causal relationship between menstruation and the incidence of anaemia is not consistent with Looker's findings that the incidence of iron-deficiency and iron-deficiency anaemia appears to vary with the race, number of children, education and poverty level of their research subjects.

Citing the work of Katharina Dalton and several others who purport to document women's impaired state during the pre-menstruum (Coutinho and Segal 68–80), Coutinho and Segal also attempt to argue that suppression of ovulation by means of menstrual suppressant contraceptives is the 'most logical rationale for resolving the discomfort of PMS at its source' (72).[8] Citing Peter Schmidt *et al.*'s study conducted on only 20 women, who showed improvement in their symptoms (which included sadness, bloating, food craving, irritability, and so on) after being put on a regime of ovulation suppressants, Coutinho and Segal argue that because menstrual suppressant contraceptives alleviate PMS symptoms in some women who suffer from the condition, all females of childbearing age ought to use menstrual suppressant birth control. This argument makes no sense because, as Coutinho and Segal themselves acknowledge, the study in question noted that the control group of women without PMS did not experience any mood changes whilst on the ovulation suppressants: it makes no sense to 'treat' someone for a condition she does not have (Coutinho and Segal 70). Their willingness to espouse such obviously weak arguments speaks of a commitment to the idea that the female body is inherently flawed. In fact, this notion appears to motivate them to argue that delaying menarche in young girls (by 'treating' them with menstrual suppressant contraceptives) might be a solution to the problem of incestuous rape:

> In Brazil, almost 30,000 pregnancies in girls age 10 and 11 were reported in 1995. Family members and friends impregnated most of these girls. A girl's biological father was often the sexual assailant. The direct solution to this problem, which should be applied forcefully and without compromise is to control the offenders or to remove them from the home. ... A late menarche is no guarantee of protection, but if (it) delays her pubertal change, the young girl may be less likely to be targeted and victimised by sexual harassment and abuse. (Coutinho and Segal 162)

Even in a case of evident criminality against young girls Coutinho and Segal seem unable to avoid locating a fault within the female body.

The Population Council

Segal occupies a key position at the Population Council, a transnational non-profit organisation with a populationist agenda. In the past he has been influential in guiding the development of various forms of contraceptive devices, including intra-uterine devices and sub-dermal hormonal contraceptive devices such as Norplant and Jadelle. Although Coutinho association with the Population Council has been more limited, he was instrumental in the development and testing of Depo-Provera (Coutinho and Segal 8–9). By virtue of his position at the Council, Segal has also been instrumental in shaping global family-planning programs by serving in an advisory capacity to the World Health Organisation, the United Nations Population Fund, the World Bank, the European Parliament, and the US Congress ('Staff' par. 2). Established in 1952, the New York-based Population Council is the brainchild of John D. Rockefeller, a member of the American Eugenics Society since 1930.

The Council claims that its mission is 'to improve the well-being and reproductive health of current and future generations and to help achieve a humane, equitable, and sustainable balance between people and resources' by developing and encouraging the use of contraceptives in poverty-stricken areas of the world with high population growth ('About the Council' par. 1). The Council characterises Rockefeller's interest in the issue of population growth as purely humanitarian, arising out of his travels through poverty-stricken and densely-populated regions of South and East Asia ('About the Council' par. 3). Thomas Shapiro indicates that a 1950 Rockefeller Foundation report, using a Malthusian analysis of the relationship between population, human welfare, social change and the threat of political upheaval, significantly shaped Rockefeller's views on population growth amongst the global poor (64). The report recommended that human fertility ought to be reduced 'so that growth can be kept at the least dangerous possible' (qtd. Shapiro 64). At its inception, on the basis of a 1952 conference of like-minded industrialists and researchers organised by Rockefeller, the trustees of the Population Council included Frederick Osborn (Advisory Council Member, American Eugenics Society, 1928–81), Frank Notestein (Director, American Eugenics Society, 1950–56) and representatives of some of the nation's wealthiest industrialist families, including Rockefeller, Mellon and Ford (Shapiro 65–6).

The Council's mission statement expressed the belief that high fertility rates were the crucial cause of widespread poverty in Africa and Asia, endangering the survival of the species and occasioning global economic and political instability. It failed to appropriately acknowledge that the poverty pervasive in the 1940s and 1950s in Asia and Africa was the product of complicated interaction between local and global multiple systems of oppression, including the systematic exploitation of the labour force and natural resources of these regions by colonisers. Working in conjunction, these factors gave few the opportunity to escape poverty. Even today, the Council fails to acknowledge that the present cause of poverty and political instability in developing nations is attributable to a number of factors, including corrupt political regimes, interaction between local and global multiple systems of oppression, systematic exploitation of natural resources and labour by transnational corporations, unjust trade agreements benefiting global powers and transnationals, misguided developmental policies and massive military budgets. The shortcomings of the Council's agenda have not limited its ability to marshal the support of various non-governmental and governmental agencies across the globe, including, at present, the

US government, the World Bank, the United Nations, the World Health Organisation and various governmental entities.[9] In the interest of providing an account of how the Population Council came to have global influence, I focus on the history of its relationship with the US.

The Population Council and the US

Frank Notestein, as Laura Landolt notes, was instrumental in marshalling the support of the US government for the Council's populationist agenda. After travelling through China in 1949 (on a trip funded by the Rockefeller Foundation), Notestein, 'influenced by China's fall to communists. ... suggested that US efforts to promote liberal political and economic institutions might be threatened by political instability and communist expansion amongst developing nations' whose disaffected rural citizenry was mired in poverty (13–14). Prior to and in the wake of the Second World War, the US government was deeply committed to the Domino Theory about the spread of communism. Therefore, Notestein's analysis suggesting that curtailing population was the solution to preventing the spread of communism would have garnered the US administration's support for the Council. Bonnie Mass also points out that the spread of communism in resource-rich nations in Central America, Asia and Africa was seen as a serious threat by the premiere US industrialists. The nationalisation of mining and various other industries in these regions of the world would seriously impact the ability of American industries to obtain raw material. According to Mass, the President's International Development Advisory Board, chaired by Nelson Rockefeller, estimated that nearly three-quarters of the raw materials used in the US originated in Asia, Africa and Latin America (35). Given that the Council's agenda of curbing population growth could help ensure American industries access to raw material by curtailing poverty and stemming the spread of communism in Asia, Africa and Central America, it was in the interest of the US government to establish ties with the organisation. Collaboration between the two entities has flourished, translating into significant budgetary support for the Council. In 2000, the Council received $41.12 million or 58 per cent of its budget from the US. In 2001 it received $37.3 million or 53 per cent, and in 2002 $37.84 million or 51 per cent (*Population Council Annual Report 2000* par. 6; *Population Council Annual Report 2001* par. 46; *Population Council Annual Report 2002* par. 7).

The vast majority of the Council's budget currently goes into the development of new contraceptive devices and research about successful family-planning programmes for the poor. Although the threat of communism no longer substantiates a government rationale for global population control, the US continues to support the Council's work because it remains committed to the idea that curbing population growth amongst the poor is the key to global economic development and fostering democracy. In the recent past, as pharmaceutical companies have expressed reluctance to engage in contraceptive development (citing the risk of product liability lawsuits and social controversy) the US government has taken the onus of supporting – albeit indirectly – contraceptive development. Rather than providing funding for the pharmaceutical companies, it funds the Council, a 'non-profit' organisation. The development of the sub-dermal contraceptive implant, Norplant, is a case in point of this approach. The Population Council invested $23.5 million on researching and $16 million on introducing Norplant in developing nations. As the groundwork for Norplant development and marketing was already in place, Leiras Oy, a Finnish company, purchased the license to manufacture Norplant from the Council and spent an estimated $23 million in developing the manufacturing protocol. Because of the Council's efforts, Norplant was picked up by the pharmaceutical company Wyeth-Ayerst, which introduced it in the American private sector at a cost of $50 million. As the pharmaceutical giant went on to garner significant profits from a product developed mostly through public funding, it is unclear whether the claim of pharmaceutical companies that it is never profitable for them to enter the contraceptive market is justified. For example, after being introduced in the US in February 1991, one million women adopted Norplant during its first full year on the market in the US, its sales totalling $141 million. This meant 800 devices being inserted in women each day (*Contraceptive Research* 108–10).

Although a class action lawsuit was initiated by US users in 1999 because the company had downplayed the side-effects of the product, Wyeth-Ayerst's parent company, American Home Products Corporation, settled the case to its advantage, making a payment of about only $50 million to approximately 36,000 women. Litigation analysts have argued that this settlement did not pose much of a burden to the company given its net profit of $2.5 billion in 1998 – no small amount of which was attributable to the global sales of Norplant (Ornstein par. 23).[10] All in all, a number of pharmaceutical companies found it in their interest to participate in Norplant manufacture and sales because of the relationship between the Council and the US. That relationship has also translated into a symbiotic bond between the Council and the US Agency for

International Development (USAID). Using funds provided by various governments, non-governmental organisations and private individuals, USAID decides which contraceptive devices should be purchased for supply to the family-planning clinics for the poor in developing nations (Hartmann 256). It also funds and directs the training of the staff for these clinics (US Agency 6). The substantial financial support that USAID provides for the Council's contraceptive development efforts and family-planning program research translates into products that USAID uses to meet its own foundational goals ('Evaluation' vii). The development of Norplant and Depo-Provera as well as their use in family-planning clinics for the poor globally, for instance, are the fruit of their co-operation.

The Population Council and the future of menstrual suppressant contraceptives

Given the Population Council's history of influence with various governmental and transnational financial institutions, if the Council throws its full weight behind Coutinho and Segal's argument for the use of contraceptives that suppress menstruation, it is not inconceivable that – in the name of promoting women's health – these organisations could influence the governments of developing nations to 'encourage' the widespread use of such contraceptives by their female populations by means of family-planning programmes for the poor. If there is such a push, the financial benefits that would accrue for pharmaceutical companies and for the Council, which owns the patent and licenses for various contraceptive devices that serve as menstrual suppressants, are worth taking into account.[11] Whether the use of menstrual suppressant contraceptives universally translates into better health for women is arguable. In *No More Periods?*, as part of her critique of Coutinho and Segal's work, Susan Rako carefully documents the relationship between the use of hormonal contraceptives that suppress menstruation and the increased vulnerability to the human papilloma virus infection and cervical cancer, along with the other risks associated with hormonal birth control devices (100–39).

Although the decision of women who, under no pressure from poverty, make informed choices about using hormonal contraceptives must be respected, are women in developing nations provided with information that would allow them to make informed choices, under conditions in which their decisions are not dictated by poverty? In this regard, it is instructive to look at USAID/World Bank/Population Council-supported family-planning clinics for the poor in Bangladesh during the 1980s. Betsy Hartmann has documented the abuses and problems with these programmes (208–26). Working in conjunction with the local government, these family-planning clinics used food and other material incentives to 'convince' a desperately poor population to use birth control, both reversible and permanent. The female clientele were routinely not provided with adequate information that would allow them to make informed decisions about contraceptive use. Nor were these women provided with the medical follow-up and support required for the safe use of various forms of contraceptives.[12] This is not to characterise the women of developing nations as helpless victims, lacking any agency to speak out against the violation of their human rights. Local grassroots feminist organisations have successfully organised movements against abuses perpetrated by these programmes.[13] But it would be unwise to underestimate the scope of influence of organisations such as the Population Council, the World Bank and USAID, especially when they identify their policies, decisions and projects as motivated by concern for women's health.

In recent years, all three of these entities have been vocal in their opposition to coercive practices in family-planning clinics for the poor, suggesting genuine concern for the plight of women in poverty-stricken nations. Whether that commitment has translated into practice is unclear. A 1995 publication of the INFO Project (a USAID-funded project) advises family-planning programmes for the poor in developing nations to be less than forthright in discussing possible side effects. According to the report,

> Because bleeding changes and weight gain are so common, during counselling all women who choose injectables should be told of these likely changes. Program managers need to decide what other side effects to mention based in part on the side effects most often reported by clients. These decisions should be made with the goal of helping clients to make a fully informed choice and to use the method effectively and confidently. ('New Era' par. 2)

Recommending that programme managers decide which side-effects to mention based on whether or not it is perceivable to the user allows for the possibility that programme managers may consider themselves at liberty to not mention the risk of cancer and stroke.

Medicalisation of menstruation

In recent years, in the name of women's health, the US has seen two distinct attempts to medicalise the monthly occurrence of menstruation.

Neither the theories of *Williams Obstetrics* nor those of Coutinho and Segal can be charted with any degree of precision using gender as the sole axis of analysis. For that reason, this chapter has tracked the complex of factors motivating and shaping the efforts to construe menstruation as an ailment. The intent of the chapter has been two-fold: firstly, to bring to light the complicated interplay between the multiple factors producing and shaping the needless medicalisation of species-typical aspects of women's reproductive physiology; and, secondly, to demonstrate that more than a willingness to ascribe the pathologising of women's physiology to a patriarchal worldview is needed to comprehend the 'truths' which these theories purvey.

Notes

1. These ideas, in one form or another, are also voiced in the eighteenth (1989), twentieth (1997) and twenty-first (2001) editions of this textbook.
2. See Martin for a developed critique of Dalton's research (*passim*).
3. In these poorer nations, Short argued, children were 'looked upon by parents as the best insurance against old age, and by politicians as a way of ensuring that their country's voice would be heard in the council of world' ('Reproduction' 124).
4. I use the term 'populationist' to refer to the school of thought that holds that high population growth is the most critical threat to economic development and democracy.
5. They make no mention of economic exploitation, sexism, racism, industrial pollution or militarisation as significant threats to human well-being.
6. According to the National Institute of Allergy and Infectious Diseases (NIAID),

   [h]ookworm eggs are passed in human feces onto the ground where they develop into infective larvae. When the soil is cool, the worms crawl to the nearest moist area and extend their bodies into the air. They remain there – waving their bodies to and fro – until they come into contact with the skin, usually on a bare foot, or until they are driven back down by the heat. Hookworm is widespread in those tropical and subtropical countries in which people defecate on the ground and soil moisture is most favorable. ('Parasitic' par. 23)

   I owe this point to Dr I. Khan.
7. By means of an extensive search of the JAMA archives on the subject of anaemia, I was able to identify the article in question. Coutinho and Segal failed to cite the authors, the title or the date of publication of the article, as they also do with a number of their cited studies.
8. In arguing this, Coutinho and Segal repeatedly fail to adequately cite their references (68, 74). These assumptions of blind faith on the part of their readers reflect their assurance in the justification of their own epistemic authority, not an approach compatible with the conception of women as rational agents, interested in and capable of making informed health decisions.

9. The staffers of the Population Council routinely serve in advisory and consultant capacity to the World Bank and other organisations (*Population Council Annual Report 2001* 51). It may well be because of the influence of the Council that the World Bank has introduced lowered population growth as a lending term on its loans to developing nations.

10. It is worth asking whether any sort of legal redress is available to women who access Norplant and other contraceptives through family-planning clinics for the poor in Third World countries but were not provided with complete information about these products. This possibility significantly impacts the profit margins of pharmaceutical companies.

11. The Council would argue that as a non-profit organisation it is not motivated by the possibility of financial gain but by its mission 'to improve the well-being and reproductive health of current and future generations and to help achieve a humane, equitable, and sustainable balance between people and resources' ('About the Council' par. 1).

12. For an account of the human rights abuses perpetrated in various nations under the auspice of family-planning for the poor, see Hartmann (168–249) and Mass (201–12).

13. Hartmann identifies various other grassroots feminist organisations in Third World countries that have spoken out against human rights abuses by family-planning programs (Hartmann 171–2, 286–96).

Works cited

'About the Council.' Population Council. 2 June 2004. 16 April 2004. <http://www.popcouncil.org/about/about.html>.

Contraceptive Research, Introduction, and Use: Lessons from Norplant. Ed. Polly Harrison and Allan Rosenfield. Washington: National Academy Press, 1998.

Coutinho, Elsimar and Sheldon Segal. *Is Menstruation Obsolete?* Oxford: Oxford University Press, 1999.

Dalton, Katharina. 'Effect of Menstruation on (British Boarding) Schoolgirls' Weekly Work'. *British Medical Journal* (30 January 1960): 326–8.

——. 'Menstruation and Crime'. *British Medical Journal* (30 December 1961): 1752–3.

Hartmann, Betsy. *Reproductive Rights and Wrongs: The Global Politics of Population Control and Contraceptive Choice.* New York: Harper & Row, 1987.

Kistner's Gynecology & Women's Health. Ed. Kenneth J. Ryan *et al.* St Louis: Mosby, 1999.

Landolt, Laura. 'Egypt and the Politics of Population Control.' PhD. Diss. University of Arizona, 2004.

Looker, A.C. *et al.* 'Prevalence of Iron Deficiency in the United States'. *JAMA* 277.12 (26 March 1997): 973–7.

Martin, Emily. *The Woman in the Body: A Cultural Analysis of Reproduction.* Boston: Beacon, 1992.

Mass, Bonnie. *Population Target: The Political Economy of Population Control in Latin America.* Brampton: Charters, 1976.

'New Era for Injectables'. *Population Reports* 23.2. August 1995. *Infoforhealth.Org.* 20 April 2004. <www.infoforhealth.org/pr/k5/k5chap4.shtml#top>.

Ornstein, Charles. 'Norplant Company Agrees to Settle Suits: Offer to Women Could Top $50 Million'. *Dallas Morning News* (26 August 1999). 18 April 2004. <www.legalactionforwomen.org/norplantsettlement.html>.

Overton, Elizabeth. 'United Kingdom by Population Groups'. *Population Change and Social Planning: Social and Economic Implications of the Recent Decline in Fertility in the United Kingdom and the Federal Republic of Germany*. Ed. D. Eversley and K. Wolfgang. London: Edward Arnold, 1982. 20–61.

'Parasitic Roundworm Diseases'. *Health Matters*. National Institute of Allergy and Infectious Diseases, US Department of Health and Human Services. February 2001. 22 April 2004. <www.niaid.nih.gov/factsheets/roundwor.htm>.

Population Council Annual Report 2000. Population Council. 2001. 18 April 2004. <www.popcouncil.org/about/ar00/financials.html>.

Population Council Annual Report 2001. Population Council. 2002. 18 April 2004. <www.popcouncil.org/pdfs/ar2k1.pdf>.

Population Council Annual Report 2002. Population Council. 2003. 18 April 2004. <www.popcouncil.org/about/ar02/financials.html>.

'Progress against Worms for Pennies'. *The State of the World's Children 1998: Focus on Nutrition*. UNICEF. 7 April 2004. <www.unicef.org/sowc98/panel20.htm>.

Rako, Susan. *No More Periods?: The Risks of Menstrual Suppression and Other Cutting-Edge Issues about Hormones and Women's Health*. New York: Harmony, 2003.

Schmidt, Peter *et al.* 'Differential Behavioral Effects of Gonadal Steroids in Women with and in Those without Premenstrual Symptoms'. *New England Journal of Medicine* (22 January 1998): 209–16.

Shapiro, Thomas. *Population Control Politics: Women, Sterilization and Reproductive Choice*. Philadelphia: Temple UP, 1985.

Short, R.V. 'The Evolution of Human Reproduction'. *Contraceptives of the Future: A Royal Society Discussion*. London: Royal Society, 1976. 3–24.

——. 'Reproduction and Human Society'. *Artificial Control of Reproduction*. Ed. C.R. Austin and R.V. Short. Oxford: Alden, 1972. 114–40.

'Staff Biographies'. *Population Council*. 9 June 2004. 17 August 2004. <http://www.popcouncil.org/staff/bios/ssegal.html>.

US Agency for International Development Bureau for Program and Policy Coordination. 'USAID Policy Paper: Population Assistance'. September 1982. 25 April 2004. <www.usaid.gov/our_work/global_health/pop/populat.pdf>.

Williams Obstetrics. Ed. Paul McDonald *et al.* Norwalk: 19th ed. Norwalk, Conn.: Appleton & Lange, 1993.

Part II
Myth and Culture

11
Menstrual Misogyny and Taboo: The Medusa, Vampire and the Female Stigmatic

Marie Mulvey-Roberts

> Oh! Menstruating women, thou'rt a fiend
> From which all nature should be closely screened
>
> (Pearsall 267)

During the First World Dracula Congress in Transylvania held in 1995, Sky News requested an interview on my research into vampirism and menstruation, only on the condition that I did not mention the word 'menstruation' or its euphemisms, as the report was going to be broadcast on prime-time television. The irony of this censorship was that my main point, that vampirism could be seen as a metaphor for menstrual taboo, had itself become taboo. While it was perfectly acceptable for me to talk about blood-drinkers such as Dracula, and his historical antecedents, Vlad the Impaler and the alleged mass-murderer Countess Elizabeth Bathory, any reference to 'menstruation' was forbidden.[1] Menstrual taboo is a complex and contested concept that has been both protective and oppressive to women (Buckley and Gottlieb 6–26). Within mythology, literature and history, it has been displaced onto images that invoke holy dread, horror and awe such as the Medusa, the vampire and the female stigmatic. These representations of the bleeding woman embody aspects of the divine and the degraded, the sacred and the sacrilegious, which have seeped into menstrual myth and misogyny. The Medusa can be seen to represent the denigration of the menstruation-based sacred female mysteries of pre-historic humans, which were further displaced and disparaged by the purity prohibitions of Judaism and the blood cults of Christianity, expressed through Bram Stoker's *Dracula* (1897). In turn, these were subsequently subverted by female stigmatics, whose bleeding from sacred wounds may be interpreted as an act of resistance to the derogation of the menses.

Reactions to menstruation are sometimes marked by such ambiva-
lence. For example, the power of menstrual blood as simultaneously
destructive and constructive is conveyed by the Roman naturalist Pliny
the Elder in the first century CE:

> Contact with it turns new wine sour, crops touched by it become
> barren, grafts die, seeds in gardens are dried up, the fruit of trees falls
> off, the bright surface of mirrors in which it is merely reflected is
> dimmed, the edge of steel and the gleam of ivory are dulled, hives of
> bees die, even bronze and iron are at once seized by rust, and a horrible
> smell fills the air; to taste it drives dogs mad and infects their bites
> with an incurable poison. (bk. 7, ch. 15, 64–5)

> For, in the first place, hailstorms, they say, whirlwinds, and lightning
> even, will be scared away by a woman uncovering her body while her
> monthly courses are upon her. (bk. 28, ch. 23, 77)

> [L]inen boiling in the cauldron will turn black. (bk. 28, ch. 23, 79)

> The bitumen that is found in Judaea, will yield to nothing but the
> menstrual discharge; its tenacity being overcome, as already stated,
> by the agency of a thread from a garment, which has been brought in
> contact with this fluid. (bk. 28, ch. 23, 80)

But this attribution of power suggests the place of menstruation at the
centre of a sex–gender system. According to Julia Kristeva, menstrual
blood has been perceived as dangerous by signifying 'the danger issuing
from within the identity (social or sexual); it threatens the relationship
between the sexes within a social aggregate and, through internaliza-
tion, the identity of each sex in the face of sexual difference' (71).
Kristeva also attributes the horror of menstruation to the unspeakable
and unpayable debt to the maternal body. As a harbinger of female
reproductive power, Bruno Bettelheim has argued in his anthropological
study of puberty rites, menstruation has been a source of male envy
(242–5). Sigmund Freud locates the origin of menstrual taboo as an
'organic repression' of an attraction men feel towards menstruating
women (*Civilisation* 36n).

The Medusa

Menstrual misogyny and its mythologizing is a return of the repressed
sexual desire which, according to Freud, may be experienced as
morbid dread. The antithesis of this is the veneration associated with

menstruation within certain cultures, most evident in Ancient Goddess worship. In Athens, Athena was central to a menstrual cult (Shuttle and Redgrove 248) and had originally been one of a triple female deity along with the Destroyer aspect Metis-Medusa (Walker 629). The paradigm shift from matriarchal to patriarchal theology was signalled by Athena's defection to male domination and her transformation of the beautiful Gorgon Medusa into a monster, who turned into stone anyone who looked directly upon her. This was the traitor goddess's punishment for Medusa's sexual union with Poseidon in the Temple of Athena. Religious sacrilege, possibly combined with menstrual sexual transgression, could have been aggravating factors (it is not clear whether she was raped or seduced), which would connect with a supernatural ancient menstrual taboo, that the look of a menstruating woman could turn a man into stone (Walker 629). After being decapitated by Perseus at Athena's request, the crawling, snaky, bloody, horror head of the Medusa may be seen to mark not only the danger of breaking menstrual taboo but also the demonising of menstruation itself. As Freud noted in his comments on repression and menstrual taboo, '[t]his process is repeated on another level when the gods of a superseded period of civilization turn into demons' (*Civilisation* 36n). Athena's rebirth from the head of Zeus, who was 'the embodiment of the new order', like the Perseus myth, 'may also be the recording of a decisive change in religious and political organization' (Baring and Cashford 343). As Robert Graves explains,

> Medusa was once the goddess herself, hiding behind a prophylactic Gorgon mask: a hideous face intended to warn the profane against trespassing on her Mysteries. Perseus beheads Medusa: that is, the Hellenes overran the goddess's chief shrines, stripped her priestesses of their Gorgon masks, and took possession of the sacred horses – an early representation of the goddess with a Gorgon's head and a mare's body has been found in Boeotia. (17)

The flying horse Pegasus and the warrior Chrysaor sprang forth from the drops of her blood; these same drops, when they fell on the desert, were transformed into venomous snakes by Gaia.

It is clear that the Medusa's blood was regarded as containing magical properties: 'Medusa had magic blood that could create and destroy life; thus she represented the dreaded life-and death-giving moon-blood of women' (Walker 629). Originally, the Medusa was a Libyan serpent or moon goddess or priestess. Some early myths claimed that the menarche was activated when a young girl was bitten by a snake god or goddess, living in the moon (Creed 64). The Latin derivation of the word 'menstruation' is 'moon change', while the Greek word for vampire, *sarcomenos*, translates as 'flesh made by the moon'. The Medusa's image could only be made harmless if deflected onto a reflecting (moon-like) surface. This resonates with an old superstitious belief that 'a menstruous woman doth infect a looking glasse as it were with some materiall corruption' (qtd. Opie and Tatem 247), while her supernatural equivalent, the vampire, as other, could cast no reflection in a mirror. In 'Medusa's Head', Freud argues that the male subject, on seeing the gorgon, experiences her petrifying agency as the stiffening (*starrwerden*) of horror at the sight of the female genitalia, essentially that of his mother, which results in an erect penis. According to Freud, this is the petrified male child's defensive fantasy against the fear of being castrated, slightly mitigated by the sight of the phallic snakes, which, incidentally, had menstrual associations for pre-historic humans.[2] Unlike castration, a menstrual period, which also involves bleeding from the genitals, appears almost magically to neither endanger life or normally to cause intense pain. The psychologist C.D. Daly, in his article on 'The Menstruation Complex in Literature' (1935), challenged Freud's view that the father is the source of the Oedipal castration complex, by arguing that it is menstruation, linked to the mother, whose power needs to be controlled through menstrual and incest taboos (340).

Judeo-Christian blood taboos

It is likely that the breaking of incest taboos, which can result in birth defects, was conflated with superstitions that intercourse with a menstruating woman corrupted semen, causing either temporary infertility or monstrous births (Ranke-Heinemann 21–2). Pliny blasted menstruating woman for having sexual intercourse, which he insisted would result in offspring that were stillborn, sickly, or 'afflicted with purulent blood serum' (bk. 7, ch. 15, 87). The Early Church fathers had a list of contraindications that included hunchbacked, one-eyed, malformed, feeble-minded, or club-headed children, while the sixth century Archbishop Caesarius of Arles, who was eventually made a saint, warned: 'Whoever has relations with his wife during her period will have children that are either leprous or epileptic or possessed by the Devil' (qtd. Ranke-Heinemann 22). The Babylonian Talmud decreed that intercourse with a menstruating woman exacted the same penalty as intercourse with one's mother. A woman's *katamenia* was believed to be one of ten curses meted out to Eve. Worse still, sexual intercourse with a woman during her menses could lead to the man's death. Alternatively, if she passed between two men, one of them would die.

In the fourth century St Jerome perpetuated the Talmudic *niddah* mentality by asserting: 'Nothing is so unclean as a woman in her periods; what she touches she causes to become unclean' (qtd. Walker 643).[3] During the seventh century, Theodore Bishop of Canterbury banned menstruating women from taking Holy Communion or entering a church for, like vampires, who were also the objects of a discourse of loathsomeness, they were considered too contaminated for contact with the sacred (Walker 643). Witches, who have been psychoanalytically identified with menstruating women, defied medieval churchmen by allegedly drinking communion wine made from menstrual blood (Bettelheim 244). This was a variation on the way in which women of the Ophite Gnostic Christian sect imbibed menses, symbolic of Christ's blood, as part of a heretical mass, *agape* or love-feast (Walker 640). In 1298, the Synod of Würzburg instructed men not to approach a menstruating woman (Walker 643). Even more polluting than menstrual fluid was the blood shed in childbirth, which required women to be 'cleansed' through the ritual of churching before being allowed to enter a church. In 1199, Pope Innocent III did not allow women in France, who had recently given birth, access to the church for the purposes of purification or even to attend their baby's christening (Ranke-Heinemann 26). Women dying before undergoing this ritual purification could be denied burial in churchyards (Ranke-Heinemann 25). For some, their liminal status was symbolised by the placing of their coffin just inside the door, rather than in the main body of the church.

The vampire

The vampire is an unholy procreator, who, like the menstruating woman, could not enter a church, and whose undead body, straddling the interstices of life and death, was believed to desecrate sacred burial ground. Stoker employs the imagery of the Medusa in *Dracula* with the female vampire with blood-dripping fangs personifying the male terror of being castrated by a voracious female sexuality. This compounds the fears surrounding the *vagina dentate* with anxieties concerning the breaking of menstrual taboo. When Van Helsing thwarts Lucy's attempt to tempt her fiancé, Arthur Holmwood, into her deadly embrace, her ensuing anger is of gorgonesque proportions:

> The beautiful colour became livid, the eyes seemed to throw out sparks of hell-fire, the brows were wrinkled as though the folds of the

flesh were the coils of Medusa's snakes, and the lovely, blood-stained mouth grew to an open square, as in the passion masks of the Greeks and Japanese. If ever a face meant death – if looks could kill – we saw it at that moment. (272)

Here the Medusa, who had huge teeth (Graves 239), can also be seen as a mirror for menstruation, reflecting the ghastly horror of breaking the sacred ban on sexual contact. Death could be Arthur's fate, if he succumbs to Lucy's seductive sexual invitation. Indeed, the dangerous effects of transgressing menstrual prohibition with his wife, Mina, had affected Jonathan's breathing, causing him to lose consciousness. But, unlike Jonathan, Arthur is not yet married and manages to resist violating betrothal boundaries. He pre-empts temptation by penetrating Lucy's body with his phallic stake in a triumphal assertion of male virility and escapes castration by her *vagina dentate*. Van Helsing then proceeds to plug up the menstrual hole of her 'lovely, blood-stained mouth' (272), and, Perseus-like, chops off her head. Decapitation was a fate that the vampire shared with the Medusa

The sexualised Lucy Westenra preys on young children for the purpose of propagating vampires. The hysteria from which Lucy is suffering had been linked by Victorian doctors to menstrual disorders (Mulvey-Roberts 84–5). In this novel, vampirism can be seen as a metaphor for the fetishisation of menstrual blood through religious imagery and symbolism relating to mirrors, the moon and the Medusa. Van Helsing's medical opinion that the depleted Lucy, whose blood is being drained by Dracula, needs replenishing with blood transfusions forms a parallel with the Hippocratic theory that menstruation was bound up with plethora and redundancy. Liz Lochhead mocked the vampirised Lucy's blood-loss through a menstrual discourse in her dramatisation:

Florrie: Miss Lucy, you all right? You does look pale.
Lucy: Oh ... nothing! I've got a visitor. ... Must have come in the night. ... my friend, my bloody friend.
Mina: The curse ...
Lucy: Don't you always feel ... unclean? Friend. Friend. How queer, some friend! (104–5)

In Stoker's novel, Dracula also targets Lucy's friend, Mina Harker, only this time he mingles her blood with his, which contains ptomaines, the generic name for alkaloid bodies such as cadaverine or putrescine, found in decaying corpses. This 'Vampire's baptism of blood' (414), suggestive of rebirth into a new state of being, also contains the ingredients of death. Similarly, the menstrual cycle signals both the potential and perishability of life. Apart from baptism, another Catholic sacrament parodied in Stoker's novel is Holy Communion, which commemorates the death and rebirth of Christ. In an attempt to purify Mina after her contact with Dracula, Van Helsing presses a consecrated communion wafer onto Mina's forehead, leaving an indelible imprint of 'the curse'.

The red stain can be read as the tell-tale mark of her having had sexual intercourse, while menstruating, with her husband Jonathan. An indication that the menstrual taboo has been breached is that Dracula's bite is not characterised by the twin puncture marks normally made by the vampire's elongated canines, but resembles instead 'a thin open wound' (366; cf. Bentley 30). The blood from Mina's neck (or neck of her womb) and 'where her lips have touched' (366), trickles onto the nightgown of her husband Jonathan, who is comforting her after the attack. In response to the horror of this menstrual trauma, Mina cries out 'Unclean, unclean! I must touch him or kiss him no more' and rubs 'her lips as though to cleanse them from pollution' (371). The bewildered Jonathan, at the sight of the blood, shrieks, 'In God's name what does this mean. ... What does that blood mean?' (365). Mina, however, realises that the curse has struck: in Leviticus 20:18, both man and woman would be ostracised for uncovering 'the fountain of her blood', only here Mina, on reading the signs, takes the blame entirely upon herself by declaiming, 'Unclean! Unclean! Even the Almighty shuns my polluted flesh! I must bear this mark of shame upon my forehead until the Judgement Day' (381–2). The redemptive powers of the sacrament prove powerless to wash away a menstrual transgression, once condemned by leading theologians as a mortal sin.[4] Mina insists that if necessary she be killed, preferably by her husband's 'loving hand', to free her from 'the awful thrall' (426). This death-wish, for having menstrually marred sexual intercourse, suggests that Mina believes her murder to be morally justifiable.

Incorporating the credo, 'For the Blood is the Life' (301), Stoker's portrayal of vampirism is a parody of transubstantiation in the Catholic Mass. This is the mantra of Dracula's disciple, the zoophagus Renfield, who eats live animals and drinks their blood: 'All lives! all red blood, with years of life in it' (359). Renfield awaits the coming of his messianic anti-Christ figure Dracula, whose physical appearance, by conforming to a Jewish stereotype, resonates in a menstrual reading of the novel, with the folk legend that Jewish men menstruate and drink Christian blood (Epp par. 8). Renfield and Dracula symbolise, respectively, the transition from Judaic to Christian blood rites, when the Old Testament covenant of animal blood sacrifices was replaced by the single sacrifice of Christ being impaled on the cross. His undead flesh and blood continues to replenish his followers through the Blessed Sacrament of the Eucharist. Dracula's ties, however, with pagan belief are clear through the fact that his powers are subject to the sun and moon (Martin 28). The novel also contains a travesty of the nuptial Mass in which Dracula declares Mina to be 'flesh of my flesh; blood of my blood; kin of my kin'; as well as his Eucharistic 'wine-press' (370), after mesmerising her into drinking his blood. Dracula is a Christ-like figure, who pierces women's bodies with marks on their necks. In a parody of the Resurrection, they are reborn through him into eternal life. In rising from the dead, they bypass the process of decay. The vampire-hunters reserved a greater disruption of bodily integrity for the female vampire. After her impalement by a phallic stake, Lucy is restored to her pre-menstrual purity, which indicates that being vampirised is synonymous with being sexualised. Similarly, Jonathan discovers the living corpse of Dracula at rest and notices fresh blood trickling from his mouth. The vampire body, like the body of the stigmatised saint, could remain incorruptible many years after death.

The female stigmatic

The drinking and shedding of blood in *Dracula* is part of a menstrual allegory relating to the sacred and the profane, which may be extended to the history of the female stigmatic. Although the ban preventing menstruating women from entering churches was lifted by the eleventh century, holy women were still denied the privilege of changing bread and wine into the body and blood of Christ or holding the consecrated host.[5] A radical alternative was for them to *become* the body and blood. The saintly female stigmatic could use her own body as a palimpsest upon which she mapped the places where the crucified Christ's flesh had been pierced. By bleeding from these sites, she purged the profanity associated with female blood through an appropriation of the sacred blood of Christ. Through transubstantiating her body into a living chalice, the stigmatic became the sacred vessel into which the sacrificial blood of the Lamb of God was poured. The Catholic devotion to the wounds of Christ, believed to precipitate rebirth through the body of Christ, is a symbolic displacement for the contemplation of the menstruant's 'wise wound' and the mystery of birth, while the stigmata itself is a subliminal insignia reclaiming from the mists of time the sacred status that was once accorded menstruation. Stigmatism subverted, to a certain extent, the notion that Christian women were meant to despise the uncleanliness of their own bodies, as reflected in the Rule for Anchoresses: 'Art thou not formed of foul slime? Art thou not always full of uncleanness?' (Walker 643).

It is significant that the vast majority of stigmatics have been female, and that it was not until the stigmata was manifested by women that bleeding from wounds became associated with the phenomenon.[6] The momentum for this coincided with the introduction of a new blood devotion to the Sacred Heart, in 1264, when Pope Urban IV ordered that the Feast of *Corpus Christi* be observed (Harrison 118). The penitent female body was believed to be in greater need of being pierced, penetrated and punished than its male counterpart. While some stigmatics bled at the time of their menses, they were mainly renowned for shedding blood, or bleeding more copiously, on Holy Days, particularly those to do with Christ's Passion, such as Good Friday, or during the period of Lent. By bleeding on prescribed feast days, stigmatics were conforming to the calendar of the Church Fathers. This was in contrast to the transgressive bleeding of menstruating women, thought to have been governed by the moon, which had pagan associations. The average 28-day menstrual cycle, also known as the lunar cycle, constitutes 13 months of the lunar year, as opposed to the solar calendar of 12 months. The number 13, traditionally associated with witches and bad luck, is also the number of disciples present at The Last Supper that had included Judas, the betrayer of Christ. The Last Supper is commemorated through Holy Communion, which, for some stigmatics, was the only food sustaining them for long periods of time. Such prolonged fasting led to amenorrhea, the cessation of the menses. According to their hagiographers, some of these holy women did not excrete or menstruate, which was seen as a 'special grace', and sometimes they exuded a sweet smell (Bynum 138). A rationalisation of female stigmatics has been to see them as vicarious menstruators, with blood issuing from parts of the body other than the genitals (Delaney *et al.* 250). Some medical theorists claimed that vicarious menstruation could also apply to men. This is encompassed within Thomas Laycock's theory, published in *The Lancet* (1842–43), concerning a general law of vital periodicity, which argued that haemorrhage and menstruation were governed by the sun and the moon.

Miraculous or sanctified bleeding is a relic of the time when the menstrual flow occupied a central and hallowed position within ancient matriarchal theology, before being banished to the margins as the curse of Eve.[7] The stigmata provided a revered discourse for bleeding women through elevating the status of female sanctity and granting women a sacred space, enabling them to rehabilitate their lost menstrual power. While women were seeking to empower themselves in the Church during the thirteenth century, the extreme ascetics, particularly stigmatics and so-called Holy Anorexics with their miraculous fasts, provided important role models of female piety. For female stigmatics, this meant that they were no longer subject to the involuntary emission of menstrual blood, but could choose effluvia that purified not only themselves, but also the world (Bynum 274). But there was a price to pay for such power and privilege. It was not merely that the stigmata, like periods, could be painful, but that self-mortification of the flesh was often *de rigeur* for aspiring or established stigmatics. Elizabeth of Herkenrode, for example, would re-enact the whole passion every twenty four hours by inflicting powerful blows upon herself. Her consolation was that just as the blood she shed did not belong to her, neither did her pain. The stigmata, in common with Christ's crucifixion, served to atone for the sins of the world which, for women labouring under the curse of Eve, were inseparable from the punishment of menstruation. Saintly women were wise not to lay claim to the blood issuing from their own bodies for, during the witch-trials and beyond, it was claimed that the menstrual fluid's evil effects were transferred to the woman herself (Larner 93). Some women stigmatics prayed that their stigmata would become invisible, though the pain remain, not unlike the taboo of silence surrounding dysmenorrhoea. Conversely, suffering was inflicted on women pretending that they had been stigmatised. In 1546 Margaretha Ulmer was punished, for such fakery, with life imprisonment, and by having her face pierced with a red hot pincer which left an indelible mark which, unlike the visible stigmata, could not heal or simply disappear (Vandereycken and van Deth 56). In her novel about a fake stigmatic, who smears herself with her menses, Muriel Spark asks, 'what else should a woman of imagination do with her menstrual blood?' (158). Another imaginative option was that employed by medieval wives, who would add it to their husband's food to decrease his libido. To increase it, they would substitute semen in dough kneaded by a woman's buttocks (Bynum 190). However empowering it was for female stigmatics to rewrite the menstrual script of the body, the inscription remained indomitably male, issuing out of the cult of the *imitatio Christi*, a union with Christ's agony on the cross.[8]

Menstrual demons

The diegesis of how the menarche sexualises girls is apparent from the locker-room scene in Brian De Palma's *Carrie* (1976), in which the heroine, while having a shower, takes a sensual pleasure in her body shortly before the onset of her first period. Carrie, who is subjected to the religious fanaticism of her mother, assumes the iconography of the stigmatic, particularly after being drenched in pig's blood, which parodies the blood shed by the sacrificial Lamb of God. This is the culmination of a practical joke carried out by her peer group, who had previously pelted her with tampons and sanitary towels after witnessing her menstrual bleed. Following her transformation into demonic stigmatic and sexually frustrated menstrual horror, the shamed Carrie comes into her power through her own telekinetic variation on the Gorgon's lethal look. The dialectic between the demonic and the divine, male misogyny and female empowerment is fleshed out in magical and mystical icons of the menstruating woman. To destroy the vampire, suppress the menstruating woman and to look away from the Medusa, the embodiment of dangerous looking, are all responses to the masculine fear of the female.

Notes

1. I would like to thank Gill Howie and Andrew Shail for their valuable editorial suggestions and Nigel Biggs, Bettina Bildhauer, Marion Glastonbury, Derek Hughes, Peter Rawlings and Dawn Starin for their help and advice.
2. According to a later version of the myth, there were copulating snakes on her belly. The serpent has been used as a totem of the cycles of life, death and rebirth and, throughout antiquity, has been associated with the immortal properties of menstrual blood.
3. See Ner-David, this collection.
4. These included Albertus Magnus, Thomas Aquinas and Duns Scotus (Ranke-Heinemann 22).
5. Women were not allowed to touch sacred vessels or vestments or to sing in the choir for certain periods in the history of the institutionalised misogyny of the Catholic Church (Wijngaards par. 17).
6. The marks that St Francis bore in 1224 did not bleed but resembled skin abrasions, whereas St Christina of Stommeln reputedly bled every Easter from wounds that had appeared in 1268 (Harrison 26–8).
7. This attitude mollified somewhat once it was agreed by some theologians that the Virgin Mary menstruated (Bynum 239).
8. This reached its height of religious fervour among women of the thirteenth century. Many eroticised this devotion, such as St Catherine of Siena who identified Christ as her bridegroom. She wore his wedding ring, which was invisible to everyone else, and died aged 33, at the same age as did her 'Saviour'.

Works cited

Baring, Anne and Jules Cashford. *The Myth of the Goddess: Evolution of an Image.* London: Arkana, 1993.

Bentley, C.F. 'The Monster in the Bedroom: Sexual Symbolism in Bram Stoker's *Dracula'*. *Literature and Psychology* 22.1 (1972): 27–34.

Bettelheim, Bruno. *Symbolic Wounds: Puberty Rites and the Envious Male*. London: Thames & Hudson, 1955.

Buckley, Thomas and Alma Gottlieb. 'A Critical Appraisal of Theories of Menstrual Symbolism'. *Blood Magic: The Anthropology of Menstruation*. Ed. Thomas Buckley and Alma Gottlieb. Berkeley: California University Press, 1988. 1–54.

Bynum Caroline. *Holy Feast and Holy Fast: The Religious Significance of Food to Medieval Women*. Berkeley: California University Press, 1987.

Carrie. Dir. Brian De Palma. Perf. Sissy Spacek, Piper Laurie and Amy Irving. Red Band Films/United Artists, 1976.

Creed, Barbara. *The Monstrous-Feminine: Film, Feminism, Psychoanalysis*. London: Routledge, 1993.

Daly, C.D. 'The Menstruation Complex in Literature.' *Psychoanalytic Quarterly* 4 (1935): 307–40.

Delaney, Janice, Mary Jane Lupton and Emily Toth. *The Curse: A Cultural History of Menstruation*. 2nd ed. Urbana: Illinois University Press, 1988.

Dijkstra, Bram. *Idols of Perversity: Fantasies of Feminine Evil in Fin-de-Siècle Culture*. Oxford: Oxford University Press, 1986.

Epp, Garret P.J. 'Scattered Flesh, Scattered Thoughts'. *Nasty*. May 2001. 1 May 2004. <http://nasty.cx/archives/000395.php>.

Freud, Sigmund. *Civilisation and its Discontents*. 1930. Trans. Joan Riviere. London: Hogarth, 1975.

——. 'Medusa's Head.' 1940. Ed. James Strachey. *The Standard Edition of the Complete Psychological Works of Sigmund Freud*. 24 vols. London: Hogarth, 1981. Vol. 18. 273–4.

Graves, Robert. *Greek Myths*. London: Penguin, 1992.

Harrison, Ted. *Stigmata: A Medieval Phenomenon in a Modern Age*. New York: St Martin's, 1994.

Kristeva, Julia. *Powers of Horror: An Essay on Abjection*. Trans. Leon S. Roudiez. New York: Columbia University Press, 1982.

Larner, Christina. *Enemies of God: The Witch-Hunt in Scotland*. London: Chatto & Windus, 1981.

Lochhead, Liz. *Dracula*. Perf. 1985. London: Penguin, 1989.

Martin, Lois. 'Flesh of my Flesh; Blood of my Blood: Feminine Subtexts and Bodily Subversions in Bram Stoker's *Dracula'*. MA Diss. University of the West of England, 1997.

Mulvey-Roberts, Marie. 'Dracula and the Doctors: Bad Blood, Menstrual Taboo and the New Woman'. *Bram Stoker: History, Psychoanalysis and the Gothic*. Ed. William Hughes and Andrew Smith. London: Macmillan, 1998. 78–95.

Opie, Iona and Moira Tatem. *A Dictionary of Superstitions*. Oxford: Oxford University Press, 1989.

Pearsall, Ronald. *The Worm in the Bud: The World of Victorian Sexuality*. 1969. Harmondworth: Penguin, 1983.

Pliny the Elder. *Natural History*. 10 vols. London: Heinemann, 1940–63.

Ranke-Heinemann, Uta. *Eunuchs for the Kingdom of Heaven: Women, Sexuality, and the Catholic Church*. New York: Doubleday, 1990.

Shuttle, Penelope and Peter Redgrove. *The Wise Wound: Menstruation and Everywoman*. London: HarperCollins, 1994.

Spark, Muriel. *Aiding and Abetting*. London: Viking, 2000.

Stoker, Bram. *Dracula*. 1897. Ed. Maurice Hindle. London: Penguin, 1993.

Vandereycken, Walter and Ron van Deth. *From Fasting Saints to Anorexic Girls: The History of Self Starvation*. London: Athlone, 1994.

Walker, Barbara G. *The Woman's Encyclopaedia of Myths and Secrets*. San Franscisco: Harper & Row, 1983.

Wijngaards, John. 'Women were Considered Ritually Unclean'. *Women's Ordination Catholic Internet Library*. No date. 21 April 2004. <http://www.womanpriest.org/traditio/unclean.htm>.

12
The Swan Maiden's Flight Over Time: Rituals, Fairytales and Matriarchy

Jacqueline Thomas

In Book Eight of Homer's *Odyssey*, Demodokus recites the story of Aphrodite's adultery with Ares. In this story, after Hephaistos traps the lovers in his iron net, he demands the return of the wedding gifts he had 'poured out for [the] damned pigeon, so lovely, and so wanton' (bk. 8, 337–9). Amidst the laughter of the other gods, Poseidon comes forward, urging Hephaistos to free Ares and swearing to Hephaistos that if he is left empty-handed by Ares, he will pay the damages (135). After Hephaistos accepts this guarantee and frees them, Aphrodite flees 'to [Cyprus] … , to her meadow/and altar dim with incense. There the Graces /[bathe] and [anoint] her with golden oil – / … and let her folds of mantle fall in glory' (387–92). Though usually seen only as comic relief in the epic, the story's references to marital exchange and escape, glorious cloth, and cosmetic ritual in a waterside Other World, particularly when considered alongside Aphrodite's Iron Age/Homeric identity as a bird, link the Aphrodite mythos with a cross-cultural folkloric bird-woman figure. It also situates them both in a menstruation-focused anthropological context. Drawing on the anthropological work of Chris Knight and Annette Weiner and on works concerning the swan maiden fairy tale, Homer's joke about the goddess can be read as masking and commemorating a prehistoric matriliny. For Knight, this social order revolved around menstrual seclusion rites in which women made themselves sexually taboo until men provided ritually-hunted meat at the end of each peri-menstruum (the days just before and after the start of the menses). Independent of Knight, Weiner also finds that around the world and throughout history, woman-produced cloth has had a singularly high cultural-ritual value derived ultimately from a vestigial, matriliny-based perception of the power of women's cycles (51).[1]

Aphrodite's marriage/escape story and the mythos of her cloth-bearing attendants, the Graces, bear signs of the menstrual seclusion and emergence rituals that the theories of Knight and Weiner, taken together, position as having been basic to this matrilineal order. These ritual signs include: the tabooed congregation of female kin in the East; cloth production; ritual bathing, dressing and cosmetic preparation; marital exchange of sex and goods. In this chapter, I amplify an established originary link between Aphrodite's bird identity and a folk figure called the swan maiden. A figure with apparently prehistoric roots, the swan maiden also possesses and produces magically-charged clothing, is captured and confined by her husband, escapes, and returns to an Other World where bathing, dressing, and cosmetic rituals with the female kin of her mother group are performed. Since menstruating women's mystical, costumed identification with animals appears to have been key to signalling symbolically their sexual 'off-limits' status – that is, as members of the wrong species for human procreation – I argue that Aphrodite's and the swan maiden's magic cloth-clad bird identities further suggest their link to the pre-patrilineal matrilinies theorised by Knight and Weiner. These correspondences, along with the swan maiden's Neolithic genesis and her appearance in the folk traditions of every peopled continent, support these anthropologists' theories of a materially-based, pan-human prehistoric social order.

Moreover, since cloth is a cultural differentiator *par excellence*, Weiner's theory of its central place in different cultures' exchange systems cannot help but attend to cultural variety. And given that Knight only posits a universal, originary social order inside a limited area of Africa 90–45,000 years ago with a correspondingly limited human population, his theory does not threaten to obscure the cultural differences that emerged as modern humans migrated out of Africa (*Blood Relations* 241–6). But it is striking that while the diversity of the many versions of the swan maiden tale attests to the variety of the human experience, the stories also seem to conserve traces of a lost cultural source. Perhaps more strikingly, the swan maiden tales span not only time and space but also genre – versions range from Maori and Native American myth to the tale of 'Hasan of Bassorah' in *The Book of the Thousand Nights and a Night* to the medieval French story of the Melusine, from Henrik Ibsen's nineteenth-century drama *A Doll's House* (1879) to two recent films, *Splash* (1984) and *Desperately Seeking Susan* (1985). Such a range bespeaks a veiled yet intimate, ongoing, pan-human familiarity with her story. The story preserves signs of matrilineal order, and its graphic violence, and in some variants, inversion of key story elements chronicle the forcible patrilineal cooption of this order.

This chapter will trace the coded traces of this lost cultural tradition from the mythic ambit of Aphrodite and from the swan maiden stories, analyzing them in the light of anthropological work on women, menstruation, culture, marital exchange and fairy tales. Aphrodite and the swan maidens of the world thus stand in their stories as ritual menstruants emerging from matrilineal, menstrual seclusion rites ready for socially-regulated conjugal sex. Usually, after encountering a forceful abrogation of this social regulation – marriage by capture – they escape. Their escapes seem to serve at once as protest against patrilineal force and as hopeful re-enactment of the menstruants' periodic return to seclusion. Beyond this 'menstrual' interpretation, by overlaying a genealogy of Aphrodite onto a genealogical sketch of anthropological theory, we can understand the negative construction of patrilineal social structures. Based as they usually are on rituals both imitative of and antagonistic to birth and menstruation, such societies revolve around male-produced sacred blood.[2] By means of this second-tier analysis, this chapter explains the oddly persistent yet largely unremarked flight of the swan maiden figure over the world's literatures and knowledge systems. As a result, certain otherwise incomprehensible details of the Aphrodite/ swan maiden figure will emerge as vestiges of a matrilineal world in which women are the founders of culture rather than the culture-polluting vessels of nature they are so often taken to be. Thus framed, the worldwide ubiquity of the swan maiden tale and Greek culture's tenacious embrace of the foreign 'trouble-making' Aphrodite can both perhaps be seen as evidence for the resilience of archaic peoples' reverence for menstruation.

Aphrodite, the swan and anthropology

Historians from the time of Herodotus in the fifth century CE until late in the twentieth century were generally agreed on the Near Eastern origins of Aphrodite, arguing that both her home on Cyprus and the Levantine elements of her divinity – one being her association with birds, often with swans – made her a Greek version of the Inanna/Ishtar deities.[3] But in 1974 Deborah Boedeker argued that she was originally a descendent of the Indo-European dawn goddess (Usas) and only later took on her Near Eastern bird attributes. Soon after, anthropologist Paul Friedrich argued that, in addition to her earlier Indo-European origins, local pre-Greek traditions also figure into her complex divinity. Thus while Boedeker argues for Aphrodite's bird associations as foreign and – because non-Indo-European – non-originary, Friedrich sees her bird aspect as coming first from these even earlier and thus also *originary* traditions. First, he posits a Proto-Indo-European swan maiden tradition as one of these local sources, arguing that its presence in East European, Baltic, Germanic, Celtic, and Norse cultures – all, like the Greek culture, Indo-European – make it likely that it 'constituted a symbolic antecedent' of Olympian Aphrodite (30). He also cites the indigenous Old European bird-goddess, present in the Baltic region before the arrival of the Indo-Europeans, as yet another prehistoric progenitor (11). This disagreement over the origins of Aphrodite's bird qualities is resolvable if both the Near Eastern and local pre-Greek traditions are seen as having a common source in an even earlier Paleolithic matrilinear social structure. Even the birdless Usas shares symbolic and narrative elements common to the stories of the Near Eastern goddesses, Aphrodite, and the swan maiden. These shared elements should be read as mythic coding of these goddesses' descents into, and emergences from, the menstrual seclusion ritual. Because these elements closely mirror the ritual enactments of girls' initiation ceremonies, past and present, around the world – all-female rites that mark or anticipate first menses with secluded, ritual enactments of cyclical, often eastern-oriented, passages, and often involving veiling and the production of special cloth – these goddesses are mythic embodiments of these girl initiates.[4]

Precisely such cyclical, female-congregated, ritualised seclusion and emergence is suggested in the basic swan maiden tale, in which an enchanted maiden, half-swan and half-human – often from the East – is bathing in a pond with her sisters when a lone hunter espies their feather cloaks lying on the shore. He steals the cloak of the smallest and prettiest among them. Since her cloak gives her the power to fly back to her Other World home, his theft not only deprives her of her immediate freedom but also compels her to a life of domestic service to him. For, once home, he hides the cloak and, as long as it is hidden, she cannot leave. There she stays and must cook, clean, and bear children for him. The story usually ends with her finding the cloak and flying away, leaving forever her husband, her children, and her domestic duties. This basic tale has several variants. In some the maiden comes willingly on condition that she not be insulted, struck, or seen naked by her husband; in others he follows her to the Other World; sometimes he sees her and her sisters spinning or weaving by the shore; in yet other versions, her magic garment might be a gown, a cap, an apron, a belt, or a pouch.

Her animal half is sometimes that of a fish, a seal, a snake, a duck, a frog, a dove, a pigeon, a sparrow, or a goose.[5]

Although the tales' elements of alternating human/animal identity, matrilocal female congregation by water, and magical 'exchanged' cloth all touch on the main components of the theories of Knight and Weiner, neither anthropologist actually analyses any one type of fairy tale. Thus the link between the swan maiden tales and these theories will be better drawn when I frame the tales with anthropological studies of them. Primary among these is Barbara Fass Leavy, who cites the known written version of the tale, the *Rig Veda*'s late third century BCE account of 'Urvasi and Pururavas' (33–7). She also establishes the tale's cross-cultural ubiquity. Yet she finds that more striking than its antiquity and its geographic range are its remarkably constant plot elements and themes, particularly its focus on housework (14–15, 38–9). Carol Silver, in her work on fairy tales in Victorian England, explains the first folklorists' preoccupations with the swan maiden tale as the result of their anxieties over women's cultural capacities and legal and marital rights (93). She notes that after having been exposed to the speculative theories of ancient matriarchies espoused by Johann Jakob Bachofen, Friedrich Engels and Lewis Morgan, most of these proto-anthropologists theorised more conservatively that the swan maiden tales were folk-historical repositories of a prehistoric matrilineal society and of its futile Iron Age struggle with evolutionarily advanced patriarchy (94–5). One of these researchers, Edwin Sidney Hartland, based his argument that the tales were such matrilineal 'survivals' on the recurring motif in the tales of the maidens' weakness to iron – a motif apparent in Aphrodite's capture in the iron net – a substance not yet used in the late Stone Age (97).

The links between marriage practices, women's rights, and fairy tales are also integral to Marina Warner's view of the fairy tale as primarily a women's genre. Warner finds that high-born women of fairy tale-crazed, Romantic-era Europe, subject to arranged marriages and hungry for stories of women marrying for love (277–8), collected and promoted many of the tales that were often published under men's names (16–21). Central to her argument for fairy tales as women's literature is the fact that these upper class women 'ethnographers' often heard the tales from their former nannies and female servants, who had learned them from their own mothers and grandmothers (24, 33–7). Exploring the stories and animal/bird associations of old wise women such as the classical Sibyls, Saint Anne, and the Queen of Sheba, she argues that these characters were progenitors of Old Mother Goose, herself a water-bird-identified character. Juxtaposing two well-known images – a 1913 children's book illustration of Mother Goose flying on a giant goose and a Classical Greek image (c. 460 BCE) of Aphrodite riding one – Warner links the two characters in her discussion of the 'knowledgeable' old tellers of the originally lewd fairy tales (51). She also views the tales' content as revealing 'women's complaint[s] and stratagems of redress', citing a swan maiden-sounding story of a stork that 'seems to threaten a husband that (his wife) will run away and refuse to come back unless things change at home' (51–3). In thus locating learned women tellers within narratives of female dignity and within the folk and revolutionary cultures she studies, Warner positions fairy tales as a cross-cultural and cross-historical tool of female resistance.

Leavy also sees the fairy tale as a form of women's protest, but interprets the swan maiden tale as a 'fantasy of options' entertained through the ages by perpetually optionless women (57). She also reads some variants' veiled and explicit references to the maidens' menstruation in the light of her view of the swan maiden as a marker for women's bodily 'naturalness' and justified inability to adapt to biophobic, male-originated culture (37, 98). And although she nods to the Victorian theories of the tales as remnants of prehistoric matrilinies, her argument depends ultimately on her view of culture as masculine oppression. But just as the Victorians' concern with their own evolutionary superiority over women and Warner's focus on the political and historical implications of fairy tales necessarily circumscribe their views, so Leavy's point of view limits hers. It draws her away from the anthropological observation that is mostly ignored in all of these analyses – namely that marriage, whether matrilineal or patrilineal, seems always to have functioned as an exchange. The swan maiden story's implicit complaint against patrilineal marital exchange, its apparent reassertion of an ersatz matrilineal system, and its cross-cultural, cross-generic presence all seem to have accidentally made it something of a nexus of 'high' and 'low' cultural deliberation. This connective reach is evident in that the tale and its 'folk' concerns about freedom, love, and survival hover over the abstracted works of early- and mid-twentieth-century anthropologists as well as those of Knight and Weiner.

Anthropology, marital exchange and patrilineage

Spurred by the post-Victorian debate over gift-giving between the sociologist Marcel Mauss and the anthropologist Bronislaw Malinowski, Claude Lévi-Strauss was the first anthropologist to fully articulate the significance of marriage as exchange (50–1, 62–8). Using field data on the ritual gift-giving of Oceanic islanders, both Mauss and Malinowski saw a universal impulse for reciprocal exchange at the heart of human culture. Yet they differed over its source: Malinowski located it externally in custom, while Mauss posited an inlying 'spirit' in gifts that was naturally felt to demand reciprocation. They argued most over Malinowski's special category of gifts that seemed to demand no return such as a man's gifts to his wife and children. Malinowski called them 'pure' or disinterested gifts, whereas Mauss saw them as amounting to a salary for a wife's sexual services (71). Lévi-Strauss expanded on Mauss's view, declaring that wives were not compelled to return these gifts because they themselves were gifts exchanged between men (Weiner 17–28). This idea was later central to his theory of elementary structures by which he explained culture as reflecting an *a priori* binary mental structure, the first instantiation of which was tribal men's perception of themselves as distinct from other groups of men. From this basic perception, he inferred, earliest men were moved to create trusting, reciprocal ties by trading their respective women. This exchange then inaugurated the incest taboo and was thus the first moment of culture. While Lévi-Strauss's views were largely abandoned by century's end, his evidence-heavy pronouncement on men's cultural exchange of women held enough currency that, by the 1990s, Knight and Weiner each felt a need to revisit it.

Although Knight considers Lévi-Strauss's use of binary mythic and social structures to be circular, because they are at taken at once both to prove and to express the binary structure of the mind, he does accept that elementary structures could exist. Indeed, he finds in menstruation a material rather than metaphysical structurer of social and mythic systems. Specifically, he sees a lasting exchange contract as having been occasioned by prehistoric women's ability to synchronise their menstruation and ovulation. Citing a 1971 study on the menstrual entrainment of spatially proximate women, Knight posits that when earliest women were pushed into close quarters near the riverine East African Rift during the Pleistocene drought 90–45,000 years ago, their menstruation synchronised. The resulting synchronisation of their ovulation, he contends, made it hard for alpha males to enforce the primate 'harem' system that had marginalised less dominant males. Knight argues that women then collectively instituted a menstrual sex strike, the breaking of which required hunted meat from the less dominant men. This strike was effected by women congregating with their mothers and sisters in ritual – often veiled and/or immured – seclusion from the hunting men during the synchronised peri-menstruum. This matrilineal power base allowed the women to leverage reliable child-provisioning help from the cooperative, less dominant men – key to women's genetic strategy, owing to the labour-intensity of feeding themselves and their children – and to block the rise of new alpha males. He further argues that they used the moon as a clock to ensure the strike's widespread synchrony and solidarity. In this way, he says, menstruation acted as something of a pan-human source of law and order, ensuring relatively fair and reliable distribution of food and sex and thereby eliminating conflict over the two 'goods' that animals necessarily fight over. The sustained peace established by this basic cultural form then allowed for more complex, symbolic cultural forms to emerge and persist. Knight thus challenges Lévi-Strauss by attributing to women the initiation of sustained political solidarity and contractual exchange, the first cultural acts (*Blood Relations* 1–49).

Knight sees this matrilineal system as having had two spheres of activity following the phases of the moon – menstrual seclusion with kin during the two-week waxing 'production' phase and ovulatory marriage during the two-week waning 'consumption' period. Under this system, a woman has two identities: during the seclusion phase she is a sexually unavailable, ritually unattractive sister or mother and, during the marriage phase she is an attractive, sexually available wife. Analysing an Arapaho myth cited by Lévi-Strauss, 'The Wives of the Sun and the Moon' (a swan maiden variant, as it happens, that explicitly deals with menstruation and marital structures), Knight finds in its collapse of the two kinds of woman into one woman under the roof of the two planetary brothers' paternal home an attempt to normalise the shift from matriliny to patriliny. Noting its patrilineal point of view, he sees a remedial apology for the cosmic disruption the men create by their conflation of the ugly, sun/marriage-resisting frog woman with the beautiful, compliant human woman whom they must teach to menstruate (Knight, 'Wives' 144–151). Reading such myths, Knight decries their theft of menstrual ritual power, their claims at cosmological authenticity by virtue of it, and their resulting pretence at being the legitimising basis for women's subordination. Knight has created a schema which polarises the two 'worlds' of menstrual matriliny. At one pole he places menstruation as symbolic death, female animality, darkness (as at the new moon), water, silence, stillness, hunting, and gender segregation; at the opposite pole, he locates ovulation, (re)birth, female humanity, (sun)light, fire, noise and dancing, feasting, and gender integration (Knight, 'Wives' 135–6; *Blood Relations* 463–6). He calls this schema the 'time resistant syntax of ritual and myth' ('Wives' 136). This logic operates in binary swan maiden motifs, principle among which are alternating female animality and humanity, gender segregation by water and gender integration by the cooking fire, menstrual ritual (symbolic death) in the Other World and ovulatory generation (birth) in the mortal world.

While Knight appears to see production during seclusion time only in terms of male meat production, I would suggest – drawing on Weiner's work on women, cloth, and exchange – that women's cloth production was a seclusion activity. Weiner also sees women as culture-producing exchangers and reveals the overlooked ritual-exchange value of woman-produced cloth in most aboriginal, traditional, and even modern cultures. She makes visible the remarkable extent to which cloth has always 'narrated' national, religious, or ethnic ideologies, cultural identity, and status (12–13). In her study of Oceanic cultures, she finds that the often supreme exchange value of woman-produced cloth derives from women's menstrual and reproductive powers. Weiner also finds that while most matrilineal and patrilineal Oceanic groups are currently patrilocal and patriarchal, the exchange traditions that Malinowski and Mauss argued over were and still are governed by matrilineal exchange rules. She notes that these economies pivot on the fact that the primary social relationship is the brother–sister bond – when a woman marries she becomes a strategist for her maternal clan, building alliances with and creating obligations from her husband's group. These ties usually involve the circulation and ritual use of the woman's textiles.

Implicit to the value of these textiles is women's standing as conduits to the world of reincarnating ancestors and their *tapu*, or dangerously sacred, status while they are menstruating, pregnant, and/or weaving (52–4). Oceanic women's crucial role in determining lineal identity, even in a patriliny, is made clear by the spiritual importance of the cloth they make. In Maori culture, for example, women are thought to imbue woven items with a surfeit of *hau*, what Mauss translated as the 'vital essence' in human beings, in land and in things' (45). It was the excessive *hau* of the Maori's most valuable items, their *taonga*, and its link to clan lineage by virtue of women's transmission that made these items desirable. Noting that some of the most sacred of Maori *taonga* are cloaks made from birds' feathers, Weiner elucidates the sacred tie between feather cloaks and their female weavers:

> The complex symbolism associated with these cloaks refers specifically to women and human and cultural reproduction. Traditional Maori cloaks made from flax fibers are called *kahu* and the same term

is used as a prefix for special cloaks, such as those made from birds'
feathers. ... But *kahu* also designates the membrane surrounding the
fetus. The term *wharekahu* is the name for the special house in which
Maori women give birth. ... Another set of meanings refers to death
and ancestors. *Kahukahu* is the germ of a human being, the spirit of
the deceased ancestor, a stillborn infant, and the cloth used by
women during menstruation. (50)

She stresses that this link between other-worldly 'cosmological phenomena' and women's bodily processes – usually facilely categorised as
profane by scholars – is for the Maori (and most Polynesians) the source
of the sacred power of these kinds of cloth *taonga* (50–2).[6]

The stable presence in swan maiden tales of feather cloaks and
other magical garments, cloth production, matrilineal congregation and
separation in and out of a watery Other World, and the occasional male
provisioning of the swan maiden's kin, all impel me to read the stories
in the light of Weiner and Knight. Interestingly, Weiner describes a Maori
baby-naming ceremony in which the newborn is placed on a fine cloak
by a stream and then dedicated to Hine-te-iwaiwa – not only the Maori
goddess of weaving and childbirth, but also a swan maiden catalogued
by the Victorian compiler, Hartland. He relates a willing-maiden variant
in which, after Hine-te-iwaiwa has killed Tinirau's two wives and married him, Hine asks him to groom her head and he says something 'dis-
respectful' about it. The insult triggers her brother's appearance in a
transporting mist, by which he rescues her. Tinirau, led by a flock of
birds, follows her to the Other World where he must first speak to her
sister before he may see her. He stays and fulfils her family's demand
that he bring them more food (Hartland 286). The distinction and conflict between the matrilineality of Hine's Other World and her husband's patrilineal harem seems evident here. In addition, Knight's
meat-for-sex contract appears as a given in Tinirau's ready compliance
with Hine's family's demand for food. In another telling, she wears
a '[flashing] scarlet-colored ornament upon her head' and emerges 'out
of the sacred [bathing] house resplendent in beauty' to attract Tinirau
(Alpers 114, 116). Her avenging brother, called her 'pigeon-brother',
takes her and her child up to the sky (113). In yet another version, Hine
convinces Tinirau to abandon his wives, in part by magically compelling
'warm clothes [to fall] at [his] feet' (Reed 147). These last three details
suggest menstrual seclusion, matrilineal bird identity, and sacred cloth
production/exchange. And, just as several myths of Hine-te-iwaiwa
taken together help to better delineate these themes, more than

one story of Aphrodite helps to trace clearly the main elements of Weiner's and Knight's theories in her divinity. Thus, while her consistent appearance as a beautiful, desirous, maiden from the East, bathed and radiantly dressed, implies that she is leaving menstrual seclusion, the Graces' intimate connection to weaving draws a fuller picture.

Aphrodite, the Graces and marital exchange

Beyond cosmetic preparation, one of the elements that Knight cites as proper to the ovulatory phase is dance, a ritual activity that Boedeker notes as proper to Aphrodite – particularly her dances with her attendants, the Graces (Boedeker 43; Knight, *Blood Relations* 345–9). Claude Calame says of Aphrodite's seduction of Anchises in the Homeric hymns that she 'passes herself off as a maiden who has not yet known love in order to seduce [him]; she disguises herself as an adolescent abducted by Hermes from [an initiatory dancing] chorus' (92). But, less discussed than Aphrodite's and the Graces' association with cosmetic and dance ritual is their link to the colour-weaving techniques used to make the cloth in Ancient Greek marital exchange. Discussing the cultural role of the Graces (*Charites*), Beate Wagner-Hasel notes their association with 'reciprocity, gratitude and social integration' (17). She explains that the Athenians regarded the Graces as wedding deities, often calling them 'well-dressed' (18–19). Making no mention of Weiner or matriliny, she does note that 'the collective character of the Graces has its roots in the collective nature of female labour and sociability' (19). In addition, her discussion of the social radiance, the *charis*, that the Graces infuse into fine cloth (and the iron weapons of warriors), resonates with Weiner's treatment of the Maori *hau* that resides in women's cloth and in warriors' spears and knives.

Wagner-Hasel gives two valences of meaning for *charis*: one implies 'reciprocation'; the other implies 'grace', a person's visually perceptible radiance (20). She combines these meanings and adds to an earlier attempt to define *charis* broadly as 'social pleasure' by 'apply[ing] the "social pleasure" to concrete subjects, to bright patterned weavings. ... Weavings are not only gifts, but also possess a *charis*ma, that can be traced back to the meaning of *charis* as a light or visual power' (20). She notes that *charis* is present in the epics in all reciprocal relationships, martial or marital. '[I]n one case, the term refers to the thankfulness of the wife, or bride, for the bridewealth given by the husband' (20). *Charis*, here, is something that can be perceived in a textile. She makes clear that the bridewealth given by husbands came under a different classification, but adds that 'the men acquire[d] the [patrilineal] right to the children as well as to the [wife's] textiles' (21). The husband's rights to this cloth gave him the kind of wealth and status that Weiner argues motivates all exchange traditions.[7] Wagner-Hasel adds that the Greek poets embraced the metaphor of narrative production as a sort of weaving that could ensure the kind of cultural continuity that came from women's colourful cloth and from the victories of the most radiant warriors (27). She says of weaving and metaphorical weaving that the Graces' 'responsibility for the visibility of the social order by means of texts, weavings and ritual performances (i.e. dances) may explain the emphasis placed by philosophers on the prominent role of the Graces in ensuring social integration, based on reciprocity' (28).

The identification of the Graces with Aphrodite, combined with Aphrodite's roots in a prehistoric, matrilineal, menstruation-ordered exchange system, makes this tradition of reciprocity-based 'social integration' comprehensible as more than a curious belief of imaginative, innately cultural Greek men. I would argue that the originary force of a larger, older, matrilineal order is strongly suggested by the fact that the charisma of the Graces' woman-made marital colour-weavings, located at the centre of the fanatically patrilineal Greek social order, can be tied to *hau*-imbued, woman-made Maori feather cloaks, to the swan maidens' feather cloaks stolen by hunters in search of wives, to the post-menstrual wedding ritual posited by Knight, to Lévi-Strauss's theory of men's trade of women, and to Weiner's articulation of cloth's power. Perceiving this unlikely web of connections, with its great span of time, place, and culture, impels me to position Aphrodite's flight home to the Graces as her periodic return to her swan maiden sisters.

Notes

1. Barber dates the advent of women's textile production at 20,000–30,000 years ago (42–70).
2. See Cook (184–5) and Jay (*passim*) on the link between male-performed ritual sacrifice and patrilineal succession.
3. See Herodotus bk. 1, par. 105.
4. See Eliade 41–7; Lamp 222–5; and Calame 103–4, 130–4.
5. See Leavy's and Hartland's chapters on swan maidens.
6. Weiner notes that Maori men at war are also temporarily *tapu* and that scholars have ignored not only Maori women's ability to attain high rank, but also warriors' capacity to pollute (52–3).
7. See Atchity and Barber on the ritual importance of Iron Age women's cloth as evidence for a waning Mycenaean matrilineal social order, slowly infiltrated by immigrant Indo-European chieftains through marriage to female Mycenaean (land) title holders (28–9).

Works cited

Alpers, Antony. *Legends of the South Seas: The World of the Polynesians Seen Through Their Myths and Legends, Poetry and Art.* New York: Thomas Y. Crowell, 1970.

Atchity, Kenneth and E.J.W. Barber. 'Greek Princes and Aegean Princesses: The Role of Women in the Homeric Poems'. *Critical Essays on Homer.* Ed. Kenneth Atchity, Ron Hogart and Doug Price. Boston: G.K. Hall, 1987. 15–36.

Bachofen, Johann Jakob. *Myth, Religion, and Mother Right: Selected Writings of J.J. Bachofen.* Trans. Ralph Manheim. Princeton: Princeton University Press, 1992.

Barber, Elizabeth Wayland. *Women's Work: the First 20,000 Years – Women, Cloth, and Society in Early Times.* New York: Norton, 1994.

Boedeker, Deborah Dickmann. *Aphrodite's Entry into Greek Epic.* Leiden: E.J. Brill, 1974.

Calame, Claude. *Choruses of Young Women in Ancient Greece: Their Morphology, Religious Role, and Social Functions.* Trans. Derek Collins and Janice Orion. New York: Rowman & Littlefield, 2001.

Cook, Erwin. *The Odyssey in Athens: Myths of Cultural Origins.* Ithaca: Cornell University Press, 1995.

Desperately Seeking Susan. Dir. Susan Seidelman. Perf. Rosanna Arquette and Madonna. MGM/United Artists, 1985.

Eliade, Mircea. *Rites and Symbols of Initiation: The Mysteries of Birth and Rebirth.* New York: Harper & Row, 1965.

Engels, Friedrich. *The Origin of the Family, Private Property, and the State.* New York: International Publishers, 1972.

Friedrich, Paul. *The Meaning of Aphrodite.* Chicago: Chicago University Press, 1978.

Hartland, Edwin Sidney. *The Science of Fairy Tales: An Inquiry into Fairy Mythology.* London, 1897.

'Hasan of Bassorah'. *The Book of the Thousand Nights and a Night.* 16 vols. Trans. Richard F. Burton. Benares: Kamashastra Society, 1885–88. Vol. 8. 7–145.

Herodotus. *The Histories.* Trans. Robin Waterfield. Oxford: Oxford University Press, 1998.

Homer. *The Odyssey.* Trans. Robert Fitzgerald. New York: Farrar, Strauss & Giroux, 1998.

Ibsen, Henrik. *A Doll's House. A Doll's House and Other Plays.* 1879. Trans. Peter Watts. London: Penguin, 1965. 145–232.

Jay, Nancy. *Throughout Your Generations Forever: Sacrifice, Religion, and Paternity.* Chicago: Chicago University Press, 1992.

Knight, Chris. *Blood Relations: Menstruation and the Origins of Culture.* New Haven: Yale University Press, 1991.

——. 'The Wives of the Sun and the Moon'. *The Journal of the Royal Anthropological Institute* 3.1 (1997): 133–53.

Lamp, Frederick. 'Heavenly Bodies: Menses, Moon, and Rituals of License among the Temne of Sierra Leone'. *Blood Magic: The Anthropology of Menstruation.* Ed. Thomas Buckley and Alma Gottlieb. Berkeley: California University Press, 1988. 210–31.

Leavy, Barbara Fass. *In Search of the Swan Maiden: A Narrative on Folklore and Gender.* New York: New York University Press, 1994.

Lévi-Strauss, Claude. *The Elementary Structures of Kinship*. Boston: Beacon, 1969.

Malinowski, Bronislaw. *Argonauts of the Western Pacific*. New York: Dutton, 1961.

Mauss, Marcel. *The Gift: Forms and Functions of Exchange in Archaic Societies*. Glencoe, Illinois: The Free Press, 1954.

Morgan, Lewis Hunt. *Ancient Society*. Ed. Leslie A. White. Cambridge: Harvard University Press, 1964.

Reed, A. W. *Treasury of Maori Folklore*. Wellington: A. H. & A. W. Reed, 1963.

Silver, Carole G. *Strange and Secret Peoples: Fairies and Victorian Consciousness*. Oxford: Oxford University Press, 1999.

Splash. Dir. Ron Howard. Perf. Tom Hanks and Daryl Hannah. Touchstone Films, 1984.

Wagner-Hasel, Beate. 'The Graces and Color Weaving'. *Women's Dress in the Ancient Greek World*. Ed. Lloyd Llewellyn-Jones. Swansea: Classical Press of Wales, 2002. 17–32.

Warner, Marina. *From the Beast to the Blonde: On Fairy Tales and Their Tellers*. New York: Farrar, Strauss & Giroux, 1994.

Weiner, Annette B. *Inalienable Possessions: The Paradox of Keeping-While-Giving*. Berkeley: California University Press, 1992.

13
Menstruating Women/ Menstruating Goddesses: Sites of Sacred Power in South India

Dianne E. Jenett

Poetry written two millennia ago in the areas now known as Tamil Nadu and Kerala in southern India described women as filled with *ananku*,[1] a sacred power associated with their sexuality that was considered particularly potent during menarche and menstruation. This Sangam era description of *ananku* is a precursor of the later concept of *shakti* (divine vivifying female power). The connection between divinity and menstruation is shown both in fieldwork and through ethnographic analyses of literature in Kerala in which the pan-Kerala goddess Bhagavati's rituals appear patterned on those of menstrual maidens. In some communities, during menarche rituals, the menstruant is understood to *be* the goddess. In this chapter I interrogate the connection between menstruation, divinity and other women-centred rituals in the matrilineal castes of Kerala. In doing so, I examine the idea that sacred power is contained in, and can be accessed through, women's bodies, and suggest how this can provide an alternative approach to menstruation from that found in Western cultures ancient and modern.

Sangam concepts

Sangam literature, written between 100 and 500 CE, reflects the worldview of an indigenous culture prior to the large-scale incursion of patriarchal Brahminic culture, which, based on Sanskrit texts, became the religion of the ruling class in Kerala during the eighth century. In the pre-Brahmin worldview, often referred to as Dravidian, the divine is experienced as imminent, within earthly reality. Ritual practices in which 'the divine is felt to be present' are sensual and ecstatic and 'the psychology of religious awareness is female' (Ram 64). The bodies of women in ancient Tamil Nadu/Kerala were considered particularly potent and full of *ananku* at menarche, during menstruation, and after childbirth. A foster mother, speaking to her daughter at her menarche, says: 'Your breasts are budding, your sharp teeth shine, your head has a coil on it, and you wear a cool dress [*talai*, a dress of leaves worn by girls after puberty].' But the daughter was now considered vulnerable: 'Do not go anywhere with your wandering girl friends – our ancient city Mutupati has [places] with attaching deities. You are protected and should not go outside' (Hart, 'Woman and the Sacred' 234). Women who had reached puberty were filled with the sacred power to affect men. This power was thought to reside in breasts, loins, and other parts of her body that made her particularly attractive (238). Sangam poetry reveals two sources of the sacred: the king and women (Hart, 'Woman and the Sacred' 233). *Ananku* is a word used to describe the powers associated with women's sexuality and women's blood which were consistent with, and equivalent to, the divine power in gods, goddesses, forces of nature, animals, warriors and kings. As will be seen later, *ananku* is a term also used for a goddess who manifests in many aspects. In two different poems, the female deity, Ananku, is said to be immanent in a woman's breasts (Hart, 'Woman and the Sacred' 239). Women and their bodies were also understood to have generative, healing, and protective powers. Although none of the songs survived, commentaries accompanying the Sangam poems mention songs women used to sing while drawing water, husking paddy, and transplanting seedlings. They apparently sang to charm the spirit of the growing plant and to coax it along as it reached fruition. They also apparently sang to soothe the pain of men who were mauled by animals or injured in wars (Tharu and Lalita 71). Wives of injured soldiers kept guard at their husband's side, to ward off evil spirits (Uyl 57).

George L. Hart III, scholar of Sangam literature, inferred that *ananku* was so 'dangerous' that menstruating women had to be tabooed. While acknowledging the powers' positive aspects, he dwells at length on the capacity of women's *ananku* to 'afflict'.

> Sacred power (*ananku*) clings to a woman and, as long as it is under control, lends to her life and to that of her husband auspiciousness and sacred correctness. But it is a power that must be kept firmly under control, lest it wreak havoc. Thus women must carefully observe chastity; they must restrain themselves in all situations' (Hart, 'Ancient Tamil Literature' 47).

Hart states that *karpu*, a term translated as 'chastity' and defined by him as restraint of 'immodest impulses' or 'marital fidelity', is the source of *ananku* (Hart, 'Woman and the Sacred' 243). V. S. Rajam, however, refutes the claim that chastity was the source of *ananku*. She cites examples in the Sangam literature of 'immodest' women described as 'an *ananku*' and unmarried women engaged in a 'clandestine, premarital love relationship with a man' as having *ananku* (263). Rajam concludes by asking, 'how is it that in ancient Tamil society *ananku* was perceived in an unmarried woman engaged in a clandestine premarital love relationship with a man? The answer is that the chief source of a woman's *ananku* was not chastity, i.e. chastity as marital fidelity' (265). Rajam further argues that 'the very word *karpu* (traditionally interpreted as "chastity") was not restricted to women in ancient Tamil society', but was also used in other situations (265).

During the period of time when this poetry was written, South India was influenced by the arrival of Jains, Buddhists and Brahmins whose religious ideas which did not exist in the indigenous Dravidian culture: asceticism, monogamous marriage for life, caste and purity/pollution. Hart's comments about chastity appear to assume the Brahminical form of forever-wedded husband and wife. In the culture of South India, at that time, marriage ties 'were loose' and 'there was no social taboo surrounding divorce, remarriage, polygamy and concubinage' (Uyl 61). He does not take into account the preponderance of the subsequent matrilineal joint family system in Kerala and parts of Tamil Nadu which followed matrilineal descent, had very loose marriage ties, and appeared to celebrate and value the sexuality of women (de Tourreil ch. 3, 5). If women's sexuality was 'dangerous' and 'uncontrollable', and marital fidelity the only method of control, it seems unlikely that a large percentage of the Tamil/Kerala population would have subsequently arranged itself in matrilineal social systems where marriage was not the norm and sexuality was not strictly confined to a single sexual partner.

The term *ananku* is often used to describe a woman's sexuality. Women whose identity and occupation centred on their sexuality were known as *parattai*, some of whom were beautiful and accomplished dancers and musicians who had a select clientele among the ruling class (Parthasarathy 288). They appear to have had a ritual function, lived in a separate section of the city, and become associated with temples (Hart, 'Ancient Tamil Literature' 47). Later these became the female ritual specialists known as *devadasi* who controlled their powerful *ananku/shakti* on behalf of the people. Historically, a variety of ritual specialists arose, each of whom had the task of attempting to devise a means 'to control this danger from within'. Some of these were female ritual specialists (*devadasi*) attached to major temples 'whose individual female powers (*sakti*) were ritually merged with those of the great goddess (Sakti)' (Kersenboom 204). It is clear that a woman's body and sexuality, celebrated in the poetry, was a site of sacred power that could be harnessed for auspiciousness. 'The force of *ananku* is potentially malevolent but can be controlled and turned into benevolent use. Similarly, the natural urge of sexual ripeness is potentially vexing and dangerous but at the same time it brings together men and women and forms the basis of all further procreation, sustenance of life, stability and happiness if properly controlled' (Kersenboom 8). *Ananku* was the goddess's power and women were the vehicle through which it was expressed. When a woman's *ananku* was especially potent, as in menstruation and after birth, self-restraint and separation were appropriate. This distinction is very important because of the ideas about menstrual pollution and impurity later brought in by the Brahmins. 'Pollution itself, which is one of the most important factors in the working of sanskritization, appears to be a development of the ancient Dravidians' notion of being infected with immanent sacred power' (Hart, 'Ancient Tamil Literature' 59). After the Brahmins rose to power and influence in the seventh century CE, the

religious elements not identified with the Brahmins kept their dangerous properties while elements which the Brahmins espoused (many of which were indigenous) were set up in opposition and were considered to be pure. Extreme measures were taken to insulate these elements from the dangerous powers and so pollution, in its modern sense, came into being (Hart, 'Ancient Tamil Literature' 43).

In Kerala, menstrual blood was considered sacred and powerful and was avoided because of its potency.

In the early literature, it is clear that the ambivalent powers of the goddess were 'recognized to inhere in young women, who performed important ritual roles embodying and expressing these powers' (Caldwell 24). The *Cilappatikaram (The Tale of An Anklet)*, an epic poem from the fifth century CE, describes a community ritual to the goddess Korravai, also called Ananku (Parthasarathy 119–29). The central figures are women: the oracle; the young maiden who 'gets the power' of the goddess, and, the goddesses' female attendants. In this scene, set in a clearing in the forest, the female oracle 'of the thundering voice' became possessed by the spirit of the goddess Ananku (119). The hair on her body 'stood on end' as she raised her hands in the air to the wonder of the devotees gathered for the ritual. She 'stirred her feet, danced in a frenzy and spoke aloud'. They chose a young maiden to embody the goddess, 'around whose thick, short, hair they tied a bowstring like the white slip of a snake, and twisted it into a tuft on her head' (120). They added a crescent moon in form of a curved tusk from a wild boar. Female attendants followed her with offerings of dolls, parrots and other colourful birds. They praised her as Ananku and carried paints, powders, cool fragrant pastes, flowers, incense, boiled rice and other food stuffs to her shrine. Describing the maiden as the goddess Ananku, the poem continues:

> The goddess wore the silver petal of the moon
> On her head. From her split forehead blazed
> An unwinking eye: her lips were coral,
> Bright as silver her teeth, and dark
> With poison was her throat. Whirling the fiery serpent
> As a bowstring, she bent Mount Meru
> As a bow. Her breasts smothered
> Inside a bodice the venomous fangs of a snake.
> In her hand, piled with bangles, she bore
> A trident. A robe of elephantskin covered her
> And over it, as Ananku, a girdle of tigerskin (121–2).

In Kerala today, rituals with oracles[2] and offerings similar to this one are still performed at festivals to Bhagavati – the Kerala goddess who has the same ambivalent qualities and iconography as the ancient Korravai/ Ananku and is understood to be her successor (Caldwell 24). The margosa or *neem* tree was one of the habitations of Korravai (Lemercinier 29), and its leaves and oil are still associated with Bhagavati and used for healing smallpox and heat-related diseases and to cool women during menstruation and after birth (Ram 86).

The goddess and menstruation

Kerala has a wide stream of *Shakta* (goddess) worship running through most of the caste ritual practices, particularly in the matrilineal communities, each of which had a family goddess. There is a deep connection with the goddess and menstruation. Tantric rituals associated with *Shakta* practice include worship of the female menstrual flow, normally considered an extremely polluting substance by mainstream Sanskritic Hinduism. Such *Shakta* notions inform the Kerala legends, in which the goddess's body manifests as the physical earth and menstruates. Until about forty years ago the menses of *Bhumi Devi* (the earth) were celebrated in a ritual called *uccaral*, which took place at the end of the second harvest in January. The season following the harvest is a period of intense heat, which progressively rises until the monsoons arrive during the first part of June. During this time, when the red land would lie fallow and the agriculturists had time off, festivals were held that required blood sacrifice. At the beginning of this season, the *uccaral* was observed, representing the menstruation and seclusion of the Goddess. For three days nothing connected with agriculture took place as she was not to be disturbed. Granaries were closed, paddy was not sold and debts were forgiven. On the fourth day the lands were reclaimed by the landlords. 'The red earth in this hot, dry season is the visible womb of the earth goddess in the season of menstruation' (Caldwell 115).

One of the most well-known temples in Kerala is the Chengannur temple where the Bhagavati menstruates. In this temple, the *tantri* (priest) examines the goddess's petticoat for bloodstains, but this has to be confirmed by the wife of the priest who determines whether it is really menstrual blood. If the goddess is menstruating, the sanctum of the temple is kept closed for three days. On the fourth day the bathing ceremony is conducted by taking the goddess to the Pampa River in procession on a female elephant, accompanied by women holding plates of offerings and *thalapoli* (lamps). Her attendants bathe her and return her to the temple, which is then reopened. The goddess's menstrual cloth is extremely valuable, having a spirit which manifests an auspicious power in the household. The opportunity to buy the powerful cloth is booked years ahead of time and chief ministers and high officials from outside the state vie for the opportunity (Grahn 3–7).

Although South Indian menarche rituals were not documented until the nineteenth and twentieth centuries, it is probable that they are a continuation from Sangam era practices. Judy Grahn's work documents in detail the correspondence of the goddess's menstrual seclusion and other Bhagavati rituals to traditional puberty ceremonies (119–213). Menarche rituals have been described in virtually all Hindu castes and communities in Kerala and were, for many communities, the most elaborate, expensive and important celebrations. Each community had different customs but they included ululation (a trilling sound made by women) to announce the event, an examination of the first menstrual stains for divination, decorating the platform the menstruant sat upon with flowers, and a ritual seclusion of the maiden. The menstruant was fed special foods to cool and strengthen her and was attended by companions. Older women sang sexual songs and gave the menarchal maiden a ritual bath (usually with turmeric). She was often covered with red cloth, presented with new clothes and important gifts, including jewellery and ritual offerings of rice from relatives. The celebration concluded with a feast and a procession of the maiden (on elephant-back for the wealthy). Similar to rituals where the ritual specialists recognise themselves as the goddess when they see themselves in the mirror, some girls were given a metal mirror to see their own divinity. Menarchal, menstruating, and post-partum women also had a special room in the house or place outdoors where they were secluded from the rest of the family. Women were considered vulnerable to 'attaching spirits' (Grahn 191–213). One view of menstruation, according to Kersenboom, is that women accumulate so much *shakti* in their blood that it has to be drained away regularly. She quotes an Ayurvedic practitioner: 'If it were not for her monthly period, five men could not hold one woman down' (69).

Women at menarche, it was found, were seen to have the kind of auspiciousness Hart described as *ananku*. Savithiri de Tourreil, a scholar from the matrilineal Nayar community, states that, at menarche, a female embodied the Goddess – became the Goddess, in a certain sense. Each female had to be prepared, entitled, properly readied, ritually, to receive the onrush of power, to become the vessel who contained the goddess 'necessary for the proper unfolding of female auspiciousness, fertility and prosperity' (Tourreil 15). And menstruation is tied to a woman's sexuality and generative capacity. As de Tourreil comments, 'it is a well-known "rule" that directly after a woman's menstruation she has the right to expect an attentive sexual partner and this is the most appropriate time for conceptions' (5). If this is so for women, it might also be so for the goddess and the end of her menstrual seclusion would signal her time of powerful creative potential. The menstruating female is seen as a purveyor of magical, if not divine, power. She is ritually assimilated by divine females who also menstruate, including the Goddess in the Chengannur temple in Kerala and the Minakshi temple in Madurai, whose menstruation 'is an outpouring of fertility, of abundance, prosperity. Hence it is viewed as highly auspicious' (de Tourreil 14). de Tourreil argues that proscriptions against menstruating females exist not because they are 'polluting', but because they are 'too pure' and thus dangerous. 'The whole thrust and texture of menarchal ritual demonstrate unambiguously that the notorious menstrual taboos make sense only as a mechanism for the protection of both the sacred female and the non-sacred categories of persons' (23).

Grahn provides exhaustive evidence that the elaborate menarche celebrations of historic and contemporary Kerala closely parallel the rituals honouring Bhagavati in modern temples in Kerala. Some puberty rites are forty-one days, the same time period demarcated for goddess festivals in the hot season after harvest. The sacred image is removed from the protected seclusion and is taken in procession on the back of an elephant around the temple courtyard to the accompaniment of ululations. The goddess is worshipped on a *pandal*, adorned in red silk, and offered flowers and cooling substances. At Kodungallur and several other temples, sexual songs are sung to the goddess. On the last day of most festivals the goddess manifests in the body of a ritual specialist who is painted and dressed to appear as the goddess. He 'gets her power' by gazing into a mirror, recognising his transformation into the goddess by seeing his face as her own. A concluding offering is *guruti*, a red liquid made of turmeric and lime often described as a substitute for blood sacrifice, and the menstrual blood of Bhagavati, a sign of her potential to create (Grahn 261). Grahn shows that many rituals to Bhagavati are modelled on menarche rituals, that Bhagavati herself is a maiden in her most potent, sexual, auspicious menarchal moment and that the devotees 'used the rituals of menarche as methods of speaking to and even directing the powers of the cosmos' (286). She sees the dialogue between devotees and the goddess as an 'intercession timed for the moment of greatest possible potential affect of the outcome – the moment in the life of the maiden when she has full (menstrual) power to create and has not yet done so' (287). She suggests that the 'construction of the goddess as a virgin maiden captures for human use' the power of Bhagavati. 'This moment is ... the cosmic creation principle ... the force of the universe that is creative of all forms, including those that have not yet occurred, and including those that are destructive'. Furthermore, '[e]ngaging this moment of all-potential in the goddess' life cycle contains within it the possibility of influencing the outcome of the formation of reality' (288). This dialogue of intercession occurs with ritual offerings.

The employment of *shakti*

In Kerala, there is a general convergence between the Brahminical and Dravidian concepts of *shakti* as a 'surfeit of a capacity or ability to do something', particularly to create and maintain life (Freeman 306). So how can we understand how *shakti* in menstruating, sexual women and goddesses plays out in daily life? Addressing this issue, Grahn explains that

> the onset of menstruation brings to a woman's body an openness to *shakti*, life energy that is comprehended as intentional (therefore deity), and perhaps earlier named as *ananku*, powerful allure in vulva and breasts that can also be harmful if it is not controlled. This power when contained creates an orderly, functional and joyous world; or when it is out of control it can burn the house down. The life energy of the sacred feminine may be called other words that have been translated into *shakti* – all power – and also pollution, emanations, influences, 'possession' and 'getting the power' (286–7).

This power of the goddess is not always welcomed in Kerala. There are different approaches to controlling *shakti*. Some communities, particularly the matrilineal ones, ritually invoke *shakti*, inviting the power of the deity to manifest concretely in the body of a person who is prepared to receive it. Other groups, including the Brahmins, use various methods to keep the power from manifesting. Just as the *shakti* of the goddess is used on behalf of the people, there are rituals where the *shakti*, the potential life force of women – at its apex at menarche, but available through ritual preparation at other times – is ritually channelled by them on behalf of their lineage. An example of the differences in dealing with the 'dangerous' power of *shakti*, shown in Deborah Neff's work, can be found in two rituals the matrilineal Nayar community can perform to their ancestral serpents. These serpents, worshipped in their sacred groves, control, through their life force (*shakti*), the harmony and prosperity of the matrilineage but can also be destructive if they are displeased. One ritual, called *pampin tullal,* is based on Dravidian practices, ideologies and concepts with female Pulluva community ritual specialists and possession of women as the central forces of its efficacy.[3] A different ritual is performed by Brahmins to control and calm the *shakti* of ancestral serpents.

The *pampin tullal* ritual is performed for the primarily middle-class matrilineal castes (Nayars) whose gods in the form of serpents have the ability to affect, both positively and negatively, a whole lineage, including generations past and future (Neff 391). The ritual specialists employed from the Pulluva community, according to their tradition, acquire their *shakti* (divine life force) from their ancestors and have the inherited capacity to take on and contain the power in their bodies. The women of the Nayar matrilineage are ritually induced into a trance state to allow them to communicate the wishes of the ancestral serpents that tell the members of the family what to do to restore prosperity and harmony to the lineage. The auspicious power of a Pulluva woman is essential for the goal of the ritual, which is fertility and prosperity, because she has *shakti* (the blessing and power of the serpent) and the serpents will come to her call. The local term used for this kind of prosperity is *aisvaryam* ('deity-ness'), auspicious increase which shines upon families and events. A Nayar woman, describing this quality, said, 'All my brothers have money, order in their lives, good things, and conveniences; and they have good marriages, good jobs and good natured spouses' (Neff 216). The *shakti* is considered part of the woman's nature, her 'propensity for sacrifice and love, and her *nagashakti* (the additional *shakti* from the ancestral serpent) – the result of devotion and practice. These character-istics render a woman potentially auspicious because they contribute to the promotion of life' (Neff 212).

South Indian women's greater quantity of inner heat (*tapassu* in Malayalam and *tapas* in Sanskrit) allows them to produce children and, by extension, 'other forms of wealth and well-being' (Reynolds, qtd. Neff 211). Certain communities and snakes also possess greater quantities of *tapassu,* as do those who suffer or sacrifice for others. *Shakti* is increased through the practice of fasting, sexual abstinence, and quiet prayerful seclusion, called *vratam,* a form of self-control which also prepares Nayar women for possession. 'The dualistic emphasis on asceticism and sexuality is predicated on a belief in immanent powers accessed through the body, united in the symbol of the serpent. Both sexuality and asceticism are believed to enhance matrilineage fertility and prosperity' (Neff 214). Neff compares women in this system with the 'truly powerful South India goddesses' and sees them as 'symbolic virgins – daughters and sis-ters whose life force is channeled for the auspicious prosperity' of the matrilineage (233). Women's *shakti* is the source of auspiciousness, good health, wealth and prosperity in the matrilineal system, where the women remained within the natal *taravad.* For all practical purposes, Neff explains, these women remain 'virgin maidens', relatively free of subordination to the affinal male. Their 'chastity' is used in service to the lineage but also can be 'dangerous' because the deities are in control through the women.

I would suggest that this ritual on behalf of the matrilineage illumi-nates a further meaning of the 'dangerous sacred power' of *ananku.* 'Danger' is unpredictability, the direct engagement with the power of deity through the body of women and the unknowable nature of what might be required/acquired as a result. There is risk and high reward in this ancient method of engaging the life force on behalf of the community. We can also see that in the matrilineal context the 'dangerous' power of women is not contained by faithfulness to a husband and is based on their literal virginity but rather on their autonomy, their freedom from subordination to a single man and their own internal preparation and self-discipline. Practices during menstrual seclusion are consistent with women's *vratam* preparation (fasting or eating cooling foods, sexual abstention and quiet seclusion). From concepts in Dravidian literature two thousand years ago and from Grahn, de Toureill, and Neff's work continuing to the present we can see that within this system sacred power (*ananku/shakti*) is seen to be contained in, and accessible through, women's bodies. Menstrual taboos recognise the sacred power of the female and were instituted for reciprocal protection; when women prepare themselves, and are free to control their own bodies and sexuality, their 'dangerous' power can be evoked, contained, and used to benefit the larger community.

Notes

1. Ananku signifies female deity as well as manifestations of her power.
2. Between the fourth and seventh centuries, with the advent of the Brahminical culture, most of the oracular and ritual power moved from the female shamans to male ritual specialists (Caldwell 25).
3. *Pampin* means 'serpent' and *tullal* refers to a song which induces ecstatic trance.

Works cited

Caldwell, Sarah. *Oh Terrifying Mother: Sexuality, Violence, and Worship of the Goddess Mali*. Oxford: Oxford University Press, 1999.

Freeman, John Richardson. 'Purity and Violence: Sacred Power in the Theyyam Worship of Malabar'. PhD Diss. University of Pennsylvania, 1991.

Grahn, Judith Rae. 'Are Goddesses Metaformic Constructs? An Application of Metaformic Theory to Goddess Celebrations and Rituals in Kerala, India.' PhD Diss. California Institute of Integral Studies, 1999.

Hart, George L. 'Ancient Tamil Literature'. *Essays on South India*. Eds Burton Stein and Society for South India Studies. Honolulu: Hawaii University Press, 1975. 41–63.

——. 'Woman and the Sacred in Ancient Tamil Nadu'. *Journal of Asian Studies* 32.2 (1973): 233–50.

Kersenboom, Saskia C. *Nityasumangali: Devadasi Tradition in South India*. Delhi: Motilal Banarsidass, 1987.

Lemercinier, G. *Religion and Ideology in Kerala*. Louvain-la Neuve: Centre de Recherches Socio-Religieuses, 1993.

Neff, Deborah Lyn. *Fertility and Power in Kerala Serpent Ritual*. Madison: Wisconsin University Press, 1995.

Parthasarathy, R. *The Cilappatikaram of Ilanko Atikal: An Epic of South India*. New York: Columbia University Press, 1993.

Rajam, V.S. '*Ananku*: A Notion Semantically Reduced to Signify Female Sacred Power'. *Journal of the American Oriental Society* 106.2 (April–June 1986): 257–72.

Ram, Kalpana. *Mukkuvar Women: Gender, Hegemony, and Capitalist Transformation in a South Indian Fishing Community*. New Delhi: Kali for Women, 1992.

Tharu, Susie J., and K. Lalita. *Women Writing in India: 600 B.C. to the Present*. Oxford: Oxford University Press, 1995.

Tourreil, Savithri de. 'Nayars in a South Indian Matrix: A Study Based on Female-Centred Ritual'. PhD Diss. Concordia University, 1996.

Uyl, Marion den. *Invisible Barriers: Gender, Caste and Kinship in a Southern Indian Village*. The Hague: International Press, 1995.

14
Medieval *Responsa* Literature on *Niddah*: Perpetuations of Notions of *Tumah*

Haviva Ner-David

In Jewish religious practice, the menstruating woman is referred to as the *niddah*. Until the destruction of the First and Second Temples in 560 BCE and 70 CE, the *niddah* was restricted in two ways: she was barred from Temple worship and by extension kept apart from all foodstuffs and vessels that were used for Temple worship; and sexual intercourse with her was forbidden and punishable by *karet* (to be cut off from the nation). The first of these *niddah* restrictions was related to what is referred to in the Bible as *tumah*, and can be most approximately translated as 'ritual impurity'. Menstrual *tumah*, or *tumat niddah*, was but one form of *tumah* mentioned in a list in Leviticus 15 of such ritual impurities contracted from bodily emissions, including seminal emissions. The second, sexual, prohibition related to the *niddah* is found in a different section of Leviticus (verses 18:19 and 20:18). These verses are not in the context of ritual *tumah*, which is reversible and which carries the consequence of being barred from the Temple; rather, they are found among a list of forbidden sexual unions, all punishable by *karet*, and all of which, if violated, result in another kind of *tumah*, moral *tumah*, which is permanent, and which, if enough is accrued, results in the exile of the nation from the Land of Israel.[1] Once the Second Temple was destroyed, however, while ritual *tumah* did not disappear, it became largely irrelevant, since with no Temple standing, there were no ramifications for the contraction of ritual *tumah*. It is only in the case of the *niddah* that one's ritual impurity status is still tracked and noted, since it is this status that determines whether the woman is sexually available or not. If a woman has contracted *tumat niddah* through a flow of blood from her uterus, she is also sexually off-limits; and therefore, due to this unique sexual prohibition attached to the *niddah*, menstruation is the only case among those listed in Leviticus 15 where laws relating to *tumah* are still in effect today. The retention of the *niddah* status, and the laws and rituals surrounding her status, has resulted in the association of the *niddah*, and women in general, with *tumah*, in contrast to men. This does not reflect the theological and Jewish legal reality. In fact, all Jews today are ritually impure from *tumah* contracted from contact with corpses, since they no longer have the ashes of the Red Heifer, which is the only means of reversing this type of *tumah*. Although *tumah* has not disappeared – having merely lost its practical relevance – even the most religiously observant Jewish men ignore their *tumah* status. A more accurate description of the current reality is that religiously observant Jewish women today who practice the laws of *niddah* are actually ritually impure from bodily emissions less often than most men, since they rid themselves of this status through monthly immersion in the *mikveh*, the ritual bath.

In general, the status of Jewish women improved in the 900–1200 CE period. As Jewish men became involved in money lending and commerce, rather than in agriculture alone, Jewish women also became more active in the public sphere and in economic life. Women also began to take a more active role in religious life, mostly as their exposure to the study of Jewish texts and the performance of ritual subtly increased. Women were also granted more civil rights, as polygamy and unilateral divorce of the woman were outlawed. At the same time, however, we have no record of significant contribution from women in the religious sphere; in contrast to the Christian and Muslim worlds, this period left no religious writings from Jewish women and no names of female Jewish mystics or scholars. One reason for this may have been the social imperative that women marry. Another reason may be the increase in superstition and belief in magic – in which women (especially menstruating women) were seen as dangerous and connected with witchcraft – as well as negative attitudes about women emerging from the philosophical schools of this period. These trends, as well as a trend towards increased piety, meant that while in many areas Jewish law was moving towards improving the status of women, in the area of *niddah*, the reverse was true. In this period, the laws of *niddah* became much stricter, and notions of the *niddah* being both physically and spiritually dangerous, as well as a source and cause of sin, proliferated (Grossman 495–507). One of the many reasons for this was the perpetuation of *tumah*-related distancing practices associated with the *niddah*. These distancing practices took two forms: those related to home life and interactions with the husband, and those related to sacred worship and ritual.

Given the theological/Jewish legal reality, why are only women associated with *tumah* today? Why has *niddah* become the receptacle of all *tumah* ideology? There are various answers to this question. The phenomenon is a complex one and relates to attitudes towards menstrual blood and women in general that are not unique to Jewish culture, as well as to Judaism-specific ideologies, traditions, and religious laws. It is my intention in this chapter to address only one such aspect of this complex phenomenon, one that has received little attention in the literature devoted to *niddah*. I will demonstrate that in the post-Temple period, from about the tenth to the thirteenth century, central Jewish legal authorities reinforced folk customs that underwrite the notion that *tumat niddah* is still relevant, thus broadening the scope of *tumah* as it relates to *niddah* both conceptually and practically, and contributing to the notion that it is the *niddah* alone who is ritually impure in the post-Temple Jewish world. I will examine *responsa* literature, which is the body of writing consisting of recorded questions asked in writing of the leading Jewish legal authorities and the recorded answers sent by these authorities which emerged out of North Africa and Spain (or Sepharad) and Italy, France and Germany (or Ashkenaz).[2]

Scholars in Sepharad and Ashkenaz

The tenth-century Babylonian, Sherira ben Chanina Gaon, was asked whether in Baghdad they should uphold the customs of not sitting upon the seat or bed of a *niddah* and of preventing her from baking and cooking for the household.[3] Whereas household members would previously have avoided sitting on surfaces upon which a *niddah* sat so as to avoid contracting *tumah*, and where the *niddah* would also have refrained from cooking for the household for this reason and out of concern that she may confer her *tumah* upon foodstuffs that may then be used for Temple worship, with no Temple standing there is uncertainty whether it is still necessary to restrict the *niddah* in this way. The questioner explained that scholars had told the people that because they cannot be rid of more severe *tumot*, such as corpse *tumah*, and would therefore be ritually impure in any case, there is no reason to uphold their restrictive practices. Furthermore, these scholars argued that the Rabbis in the Talmud forbid a *niddah* from performing only three chores for her husband – taking out his bed, mixing his wine, and washing his hands and feet – and they did so because these were especially intimate acts and therefore were more likely to sexually arouse the couple and, therefore, were meant only as a means to prevent sexual intercourse, not the contraction of *tumah*.[4] Sherira Gaon's answer was that according to the letter of the law, these scholars are correct: first, there never was a prohibition against contracting *tumah*, even when there was a Temple standing (although Jews probably did avoid trying to contract *tumah* on some level, which is why women in *niddah* would have had separate vessels and sitting furniture), and second, those three restrictions mentioned in the Talmud were meant to prevent sexual intercourse with a *niddah*, not the contraction of *tumah*. Sherira Goan then went on to rule that this community should uphold their strictures out of the fear that introducing any leniencies in the area of *niddah* could lead to couples violating the sin of having sexual intercourse with a *niddah*. This response goes to the centre of this crux point of menstrual discourse. The fact that Sherira Gaon endorsed these strictures is striking, especially in light of the fact that he was such a strong advocate of upholding the Babylonian Talmud – in which it is stated that the only three restrictions upon the *niddah* in relation to her household duties are the three mentioned above – as the highest Jewish legal authority. What is fascinating is that he endorsed folk *tumah* practices only as a means to prevent the sexual prohibition from being violated, erecting a fence around the sexual prohibition. Even more than that, he did so knowing that most practitioners would have perceived these rules as being about *tumah* and not about sexual intimacy. This highly authoritative Jewish legal authority even advocated retaining folk *tumah*-related praxis around the *niddah* despite textual objection in the Babylonian Talmud.

Isaac ben Jacob Alfasi, a major Talmudic and Jewish legal scholar writing in Morocco and Spain in the eleventh century, was more explicit. The question put to him was whether one may sit on the seat of a *niddah* who has stopped bleeding, but who has not yet immersed in the *mikveh*, since she is still in her seven 'clean' days. This refers to the seven days the *niddah* is required, by rabbinic law, to count after the bleeding has stopped before she may immerse in the *mikveh*, thereby reversing her *tumah* status and becoming sexually available to her husband.[5] His answer was that as long as a *niddah* does not immerse in *mikveh* water, she is still a *niddah*, and all the relevant laws apply to her. However, since the reason one might keep away from her chair or bed is 'only so that the laws of ritual purity will not be forgotten among Israel' (*Sheelot Alfasi* 152) and since she has actually stopped bleeding and has washed in drawn water (that is, non-ritual washing), one may sit on a chair she has sat upon or lay upon a bed she has laid on, because, technically speaking, it is permissible to do so even when she is actually bleeding. Alfasi was arguing that the custom of not sitting on the chair or laying on the bed of a *niddah* is a valid custom that is designed to prevent the laws of ritual purity and impurity from being forgotten, but that the custom was not so strong as to say that one who lays on the bed of a woman who is no longer menstruating but has not yet immersed in the *mikveh* is transgressive.[6] The custom was strong enough to be in effect while she is actually bleeding, but not strong enough to be in effect in the case of a woman who is counting her seven 'clean' days and therefore has not yet immersed. Like Sherira Gaon, Alfasi admitted to the irrelevance of keeping this practice, since the Jewish laws of ritual purity and impurity no longer applied, advocating continuing the practice of not sitting on the bed of a *niddah* only so that these laws not be forgotten. Both of these major Jewish legal authorities reinforced *tumah* notions in relation to the *niddah* through the encouragement of keeping this practice alive, but both also acknowledged the practical irrelevance of *tumah* in post-Temple Jewish life.

Another important *responsum* on this subject was written by Joseph ben Meir ibn Migashin, a major spiritual leader and scholar of Spanish Jewry in the twelfth century and a student of Alfasi. He was addressing the issue of a woman without a regular menstrual period and hence, in practice, deemed to be in constant *niddah* status. The questioner asked if this warranted a divorce in the sense that her husband could avoid paying her the requisite compensation (*ketubah*) that he would normally owe her upon divorce. A *responsum* with complex legal intricacies, two lines of it are crucial. Migash stated that '[t]his situation prevents her from serving him, since she would forever be prevented from sexual relations and household duties, as is the custom in these parts'. Later, he explicates further that 'in these parts, when the woman is a *niddah*, she refrains from performing all household chores' (*Sefer Sheelot Migash* subsect. 129). Migash's *responsum* gave evidence of, at least in twelfth-century Spain, a folk custom intended to keep the woman in *niddah* from going about her usual business, despite strong Talmudic argument against it. He did not in any way criticise this practice, but rather mentioned it as a matter of course, and even used it as part of his argument to prove that the woman with an irregular period should be divorced without her *ketubah* money. He is willing to accept customs with no basis in Talmudic law that would leave this woman destitute. His *responsum* constitutes a reinforcement or even advocation of this behaviour. This is not especially surprising, considering that he was a student of Alfasi. Nevertheless, it is indicative of the general pattern of reinforcing *tumah*-related customs despite their irrelevance according to Jewish law and despite the lack of textual basis in the Talmud for their continued practice.

The *responsum* written by Maimonides, Moses ben Maimon, in the twelfth century was a reply to a question sent by Yosef ben Jaber in Baghdad, criticising Maimonides's stated position in an earlier *responsum* (*Sheelot Uteshuvot HaRambam* subsect. 114, 197–8). In this first *responsum*, Maimonides had allowed women to retain their custom of refraining from regular household chores in general during menstruation even though household chores were formally permitted, and then to return to these duties during their seven 'clean' days, refraining only from those interactions between husband and wife forbidden by Jewish law during *niddah*. Maimonides wrote that if the questioner is asking whether a woman counting her seven clean days cannot cook and knead dough, touch clothing, or walk on a mat or pathway before others, these are all allowed even during the days she actually sees blood, since the laws of *tumah* are no longer in force. He added that this is the accepted custom in most places, although in Egypt some have adopted much stricter Karaite restrictions upon a *niddah*:[7]

> But whoever wants to be lenient can be lenient, and whoever is revolted by this because of pollution, or out of a desire to add a fence in order to distance himself from the *niddah*, he should do so. But if he understands this prohibition to touch foodstuffs or drink that which a *niddah* touched, and if he distances himself from them because he thinks it is forbidden, he has stepped outside the boundaries of Rabbinic Judaism and denied the Oral Law. (*Sheelot Uteshuvot HaRambam* 588–9)

Maimonides clearly distinguished between the *tumah* aspect of *niddah* and the sexual prohibition related to *niddah*. He was clear that there is no prohibition related to contracting *tumat niddah*, and he stated explicitly that in the rabbinic laws of *niddah*, the issue of ritual impurity has been replaced with the issue of preventing sexual intercourse. In fact, he even called one who thinks there is an actual prohibition related to touching food or drink which a *niddah* touches a heretic, probably because they would be adopting Karaite customs. By allowing these customs to continue in the name of pollution, Maimonides allowed for the idea of *tumah* to remain in the consciousnesses of practitioners, acknowledging that *tumah* as far as it relates to entering the Temple was no longer a problem. He also reinforced the negative associations that were concomitant with *tumah*. In fact, he introduced the notion of the polluted menstruant as a viable model with which to replace ritual *tumah*. This is not that far from the biblical idea of *tumah* since, even if

Maimonides did, the majority of practitioners would not have recognised the fine line between *tumah* and pollution.

Rashi, Rabbi Shlomoh ben Yitzhak, in eleventh-century France, when faced with the question of whether or not vessels a woman touches should be considered *t'meim* (ritually impure) for her husband, answered that

> [v]essels that a *niddah* touches today are ritually pure with reference to her husband, since people today are ritually impure from graves, and tents that contain corpses, and from rodents, and dead animals, and corpse *tumah* … however, we are stringent upon ourselves, and we don't eat from the same plate, and we don't eat from her leftovers, and we don't sit upon things she has sat upon, and we don't take anything from her hand. And we give her separate vessels and plates, handkerchiefs and sheets, pillows and pillowcases, for her to use during her *niddah* period. And this is a good custom, because of the concern that the couple may be lead to sin out of habit. But there is no prohibition on grounds of *tumah*. (*Machzor Vitri* 605–6)[8]

This *responsum* is revealing of the customs of the times, suggesting that, in eleventh- and twelfth-century France, Jews broadened the prohibitions laid out in the Talmud, so that women were effectively isolated from their husbands and households. No less fascinating in this *responsum* is Rashi's wording. On the one hand, he was insistent that there is no *tumah* prohibition; hence, technically, as far as *tumah* is concerned, her husband may touch vessels she has touched. However, he added that the custom is to be stricter and to completely separate from a *niddah*, which means giving her separate vessels, not sitting on her seat nor bed, not passing objects to her, and so on, in order to prevent the kind of temptation that can lead to sexual intercourse. Rashi himself would avoid the passing of keys between himself and his wife when she was in *niddah*.[9] He was not articulating a complete separation between sexual prohibition and ritual purity practices; rather he was approving the application of *tumah* practices in order to prevent the violation of the sexual prohibition. Unlike the Talmud, which clearly forbids only those relations that have an overlay of sexual intimacy, Rashi is including such clearly *tumah*-related practices as sitting on a chair she has sat on or using the same vessels she has used. Despite his belief that as regards the requirements of ritual law in relation to *tumah* there was no reason for the continuation of these folk customs, he encouraged them as an aid to the sexual prohibition. In the thirteenth century, Isaac ben Moses of

Vienna quoted from this *responsum* of Rashi, saying that Rashi gives no reason for the practice of these stringencies. He then suggested that a possible reason is that the *niddah* is dangerous, although he used the word *t'meah* to describe the *niddah*'s status.[10] He then added that despite the fact that these are mere stringencies, it is praiseworthy of the women to keep them, and all stringencies that a person keeps in the area of *niddah* will bring him/her a blessing (subsect. 360). Despite the lack of basis for these practices in the Talmud and mainstream Jewish law, he displayed no reservations about encouraging their continued practice.

Fear of the *niddah* and *tumah*

Yedidyah Dinari correlates extreme twelfth-century and thirteenth-century distancing practices during *niddah* to the medieval notion that the *niddah* is dangerous. He concludes that '[i]t is not surprising that the danger of causing harm is expressed in the terminology of *tumah*. The expression, *t'meah*, relating to a menstruant, was very accepted, and therefore they used a known term, although it received a new meaning' (321). Dinari draws a complete distinction between *tumah* and danger, stating that the danger associated with the *niddah* and not her *tumah* is the reason for these distancing practices, despite the fact that it is the language of *tumah* that is used to express these feelings of fear. Similarly, Charlotte Fonrobert addresses the issue of *tumah* terminology in Tractate *Niddah* of the Babylonian Talmud. Like Dinari, she claims that the use of *tumah* terminology in relation to *niddah* in this Tractate is a 'halakhic [Jewish legal] and conceptual misnomer, since the relevant laws of *niddah* do not concern ritual purity or impurity'. Furthermore, she refers to the use of this terminology as a 'linguistic slippage' on the part of the Rabbis that really refers to the sexual prohibition, and that this 'linguistic slippage introduced by the rabbis obscures what biblically and halakhically are two different conceptual frameworks for discussing menstruation' (29). I would argue that, in fact, we should give more weight to the choice of words by the Rabbis in the Talmud and the medieval Jewish legal scholars in order to reach a deeper understanding of why specifically this *tumah* terminology is used in the place of the language of sexual prohibition. Moreover, whereas Fonrobert's model understands the two conceptual frameworks of *niddah* (the *tumah* framework and the sexual prohibition framework) as separate, I understand these two conceptual frameworks to be intertwined.

The use of *tumah* terminology in relation to *niddah* after the destruction of the Temple did not represent a mistake, but rather a meaningful act.

These legal scholars consciously applied *tumah* language, and it is important to note that by doing so they were not misrepresenting the Jewish legal reality. The *niddah* today still is ritually impure, even if the only consequence of this *tumah* is her sexual status. Where these major religious authorities could have used the language of sexual prohibition, why did they choose not to? Dinari and Fonrobert argue that *tumah* terminology of the Talmud and later *responsa* lost all of its *tumah* meaning in this context, and that the rabbis who implemented these terms did so only because these were terms that were already in the Jewish legal lexicon, or, in the case of the medieval scholars, perhaps because they were following in the footsteps of the talmudic Sages who also used this terminology loosely.[11] But *tumah* terminology cannot be separated from *tumah* ideology. It makes sense to assume that rabbinic scholars as well-versed in Jewish law as those quoted above used their language carefully, especially in this case, where there is no reason to assume they did not, since, as I explained above, *tumah* did not disappear with the Temple, but rather lost its practical application, except in the case of the *niddah*, where her *tumah* has the added ramification of causing her to be sexually off-limits as well as barred from the Temple, which was a characteristic all ritual impurities shared.[12] Why assume, then, that these rabbis had no intention of evoking *tumah* by using that word, especially since they were using it in the context of the one case where *tumah* was still practically relevant (even if only as an indication of her sexual status) – that of the *niddah*?

Dinari may be correct when he suggests that an impetus for many of these distancing laws may have been fear of the *niddah*. But fear of the *niddah* and *tumah* are not incompatible ideas. With a range of possible interpretations of the *tumah* laws of the Bible, it is likely that fear played a part in the formulation of these laws. Therefore, it is not necessary to make a total separation between *tumah* and fear of danger. *Niddah* being the only case where two types of *tumah* – ritual and moral – converge, one may even have expected notions of the *niddah* as dangerous to emerge, since the exile which results from moral *tumah* is a frightening prospect. Thus, even if many of the distancing practices mentioned in these responsa were the result of fear of the *niddah*, this does not mean that the *niddah* was seen as ritually pure. On the contrary, she was perceived as ritually impure, a concept which included the notion that she was dangerous. All of the responsa examined here suggest a perpetuation of the notion of *tumah*. If those scholars using the language did not see ritual *tumah* as relevant to the *niddah*, their use of *tumah* language perpetuated notions of the *niddah* as ritually impure, even if contemporary cultural connotations and perceptions were that of danger or pollution. If their intention was to prevent sexual arousal and not the contraction of *tumah*, the result was the perpetuation of associations between the *niddah* and *tumah* in Jewish consciousness. Moreover, even if the experience of *tumah* among most practitioners of this ritual came to be about danger and pollution, not technical ritual *tumah*, this did not detract from the problem. In fact, it may even have made it worse, since what these authorities did was perpetuate already existing negative attitudes about the *niddah* and give them a legitimate religious framework and channel. Either way, the woman was seen as being the locus of *tumah*, even if this was a reinterpretation of *tumah*.

The steps taken by these scholars in their response have had long-lasting effects. Menstruating women, and even women in general, are still perceived in the religious Jewish world to be at best more ritually impure than men, and at worst the embodiment of *tumah* itself. While the phenomenon I discuss above is not the only factor contributing to this, it is a major factor. While most *tumah*-related distancing practices determining the husband-wife relationship and the conduct of household chores are no longer part of *niddah* praxis, some do remain, among which are the practice of avoiding sitting on the bed, bench or couch of one's wife when she is in *niddah* and not eating from her dish or leftovers. I propose a careful use of *tumah* language in relation to *niddah*. Adhering to the language of sexual prohibition may of course not be a better choice, since this language also has negative connotations of sin and improper sexuality, and is just as likely to encourage misogynist attitudes as *tumah* language. This is why alongside change in praxis must go a change in ideology. It is now in the hands of women to reverse what was put in play centuries ago by refusing to abide by those few still remaining restrictions that hark back to *tumah* avoidance in order to reinterpret *tumah* so as to create a positive understanding of what it means to be a *niddah* and to replace medieval interpretations of *tumah* that include notions of pollution, danger and filth.

Notes

1. There is no sexual prohibition attached to men who contract ritual *tumah* from bodily emissions, only to women. It should also be noted that the *niddah*, who is experiencing a normal, menstrual flow, and the *zavah*, who is experiencing an abnormal flow, became conflated into one category by the fourth century. I use *niddah* to refer to both of these categories.

2. Jewish religious customs, and in some cases even laws, in Ashkenaz differed from those in Sepharad.

3. *Gaon* was the title conferred upon the Jewish legal authorities between the seventh and the eleventh century in Babylonia, which was located in what now is Iraq. This *responsum* was preserved, in part, in a book of *responsa* of *geonim* (the plural of *gaon*) of the East and West, (*Teshuvot Geonei* subsect. 44) and also in Shalom and Hanoch Albeck's edition of Avraham ben Yitzhak of Narbonne's *Sefer Haeshkol*. Ginzberg published a manuscript from the Cairo Geniza of this responsa in his *She'elot Uteshuvot HaGeonim Min HaGeniza Asher BiMitztrayim*, 206–7.

4. See *Talmud Bavli*, Ketubot 61a.

5. According to biblical law, the *niddah*'s *tumah* status is reversed after seven days (the first day being the first day of menstrual bleeding), with no need to count another seven 'clean' days after the bleeding has stopped. It is the *zavah* who must count seven 'clean' days after her bleeding has stopped. However, according to rabbinic law, women must wait seven 'clean' days after the cessation of any flow of blood from the uterus.

6. The principle of continuing a practice that has no real halakhic significance so that it will not be forgotten is found in three places in the Babylonian Talmud (*Talmud Bavli*): Eiruvin 7a and Bekhorot 18b and 27a. Two of these cases are related to Temple worship.

7. The Kairites denied the authority of the Oral Law and hence of the Talmud and were regarded as heretics by the mainstream 'Rabbanites'.

8. The version in Ehrenreich's *Sefer Ha'orah of Rabeinu Shlomo B'Rabi Yitzchak* contains some significant differences. One is that he specifies that it is the husband who is permitted to touch vessels his *niddah* wife has touched; another is that he says that it is the vessels that she used, rather than sat upon, that she must wash in water. Ehrenreich's *Sefer HaPardes* has more differences, including a warning not to touch the *niddah*'s clothing and to generally try to distance oneself from her. Another significant difference is the inclusion of the word *viyitaher* when referring to the washing of the clothing, implying that the clothing will not only become clean by washing them, but will also be ritually purified.

9. See Tosafot on Shabbat in *Talmud Bavli*, 13b.

10. Nachmanides, Rabbi Moses ben Nachman, another major medieval Jewish scholar in Spain, also connects *niddah* with danger (255–6, 387–8).

11. Ullman points out that a binary relationship – meaning one name and one sense – is not always the case in language: 'More often than not, several semantic relations are telescoped, with the result that more than one name is attached to one sense, and more than one sense to one name, within the synchronous system' (106). He calls this 'multiple meaning', writing that 'a word may retain its previous sense or senses and at the same time acquire one or several new ones' (107). This, I argue, is the case with the term *tumah*.

12. The verse in the Talmud (Lev. 18:19) warns against drawing near a woman who is in the state of '*niddat tumatah*' in order to have sexual intercourse. I am proposing a reading of this verse that presents her *tumah* as the reason for the sexual prohibition.

Works cited

Avraham ben Yitzhak of Narbonne. *Sefer Haeshkol.* Ed. Shalom and Hanoch Albeck. Jerusalem: Reuven Mass, 1935–8.

Dinari, Yedidyah. 'Minhagei Harchakot HaNiddah' ['Impurity Customs of the Menstruate Woman']. *Tarbiz* 49 (1979–80): 302–24.

Fonrobert, Charlotte Elisheva. *Menstrual Purity: Rabbinic and Christian Reconstructions of Biblical Gender.* Stanford: Stanford University Press, 2000.

Ginzberg, Louis. *Sheelot Uteshuvot Hageonim Min Hageniza Asher Bimitzrayim.* Vol. 2 of *Geonica.* 2 vols. New York: Jewish Theological Seminary of America, 1909.

Grossman, Abraham. *Hassidot Umordot: Nashim Yehudiyot B'europa Biyimei Habeinayim [Pious and Rebellious: Jewish Women in Europe in the Middle Ages].* Jerusalem: Mercaz Zalman Shazar LiToldot Yisrael, 2001.

Isaac ben Moses of Vienna. *Or Zaruah.* 1862. Brooklyn: Goldman, 1962.

Machzor Vitri LiRabeinu Simcha. Ed. Shimon HaLevi Horvitz. Jerusalem, 1961.

Nachmanides [Rabbi Moses ben Nachman]. *Commentary on the Torah.* Trans. Charles B. Chavel. New York: Shilo, 1974.

Sefer Hapardes. Ed. H. Ehrenreich. 1924. Rpt. Bnai Brak: Yahadut, 2001.

Sefer Ha'orah of Rabeinu Shlomo B'Rabi Yitzchak. Ed. H. Ehrenreich. Bnei Brak: Yahadut, 2001.

Sefer Sheelot Uteshuvot Lirabeinu Yosef Halevi eban Migash [Responsa of Rabbi Josef HaLevi ibn Migash]. Ed. Simcha Blau and Avraham Hassida. Hotzaat Machon Lev Sameach. Jerusalem: Alef, 1991.

Sheelot Uteshuvot Rabeinu Yitzhak Alfasi z'l [Responsa of Rabbi Isaac Alfasi]. Ed. Zeev Woolf Leiter. Pittsburgh: Machon HaRambam, 1954.

Sheelot Uteshuvot HaRambam [Responsa of Maimonides]. Ed. Yehoshua Blau. Jerusalem: Reuven Mass, 1986.

Simha ben Shmuel of Vitry. *Machzor Vitri LiRabeinu Simha.* Ed. Shimon Halevi Horvitz. Jerusalem, 1963.

Talmud Bavli. 20 vols. Vilna, 1865–66.

Teshuvot Geonei Mizrah Uma'arav. 1888. Ed. Yoel Miller. Vagshal: Jerusalem, 1987.

Teshuvot HaRambam. 3 vols. 1957–61. Ed. Yehoshua Blau. Jerusalem: Reuven Mass, 1986.

Ullman, Stephen. *The Principles of Semantics.* Oxford: Blackwell, 1951.

15

'Let Him Pass for a Man': The Myth of Male Menstruation in *The Merchant of Venice*

Ariane M. Balizet

Of all his plays, *The Merchant of Venice* (1594) provides the most challenging and perhaps fruitful landscape for investigating William Shakespeare's elusive constructions of gender and gender identity. While his other comedies maintain a sharp focus on interactions between men and women, they eventually reinforce the primacy of polarised gendered identity and, ultimately, the matrimonial union. Even in his tragedies, where gender plays a significant role in the execution of business outside the domestic realm, Shakespeare leaves only the options of 'sexed' and 'unsexed'. In *Macbeth* (1606), for instance, the title character sees his status as a man challenged only in opposition to the total annihilation of his identity: 'I dare do all that may become a man;/Who dares do more is none' (1.6.47–8). Lady Macbeth subscribes to this dichotomy when she asks evil spirits to 'unsex [her] here' (1.5.41). *The Merchant of Venice* stands out against the battles of the sexes resolved by Shakespeare's high comedies, in the sense that the qualities of masculinity and femininity – not the more concrete natures of men and women – are brought to the foreground. *The Merchant of Venice* centres on Antonio, the merchant of the title, and Shylock, a Jewish moneylender. As a favour to his friend Bassanio, Antonio enters into a financial bond that entitles Shylock to a pound of his flesh if he cannot repay the loan. Meanwhile, Shylock's daughter Jessica escapes their house carrying part of Shylock's fortune to marry the Christian Lorenzo. When Antonio's ships are lost, Shylock goes to the court of Venice to demand his bond, but is ruined by the stipulation that he cannot spill one drop of Antonio's blood. My condensed synopsis of this rich play highlights the moments that most pertain to *The Merchant of Venice* as a play deeply concerned with difference, deviance, and defiance. Shylock himself comes from a long and often disturbing tradition of Jewish men in the medieval and early modern imagination. During the time in which Jews did not officially exist or reside within the kingdom, the Jew emerged as a figure consistently identified and thus vilified by a relationship to blood. Whether shed, spilled, or sold, blood stains the majority of medieval and early Modern representations of the Jewry. Shylock's relationship with the myth of Jewish male menstruation both reaffirms the myth's currency at the end of the sixteenth century and foretells its exit. This myth, a bizarre conflation of medieval stereotypes of Judaism, shows itself at key moments in which Shylock's gender is questioned. Although I am deeply indebted to several excellent studies by David Katz and James Shapiro – among others – on the purchase this myth holds in Shakespeare's play, Shylock's place in the tradition of Jewish male menstruation extends far beyond 'if you prick us do we not bleed?' (3.1.54) and similar meditations on Shylock's body. Shakespeare's invocation of Jewish male menstruation predicts Shylock's domestic, economic, and religious fate; indeed, it *is* the story of Shylock.

'The blood out-cryeth on your cursed deed': representations of Jews before Shylock

The assertion that Jews did not 'exist' in England at the time Shakespeare was writing *Merchant of Venice* in the sixteenth century is in itself a dangerous one; as Shapiro has argued, the accepted story of the Jews' expulsion and absence from England, along with their eventual readmission, is a reductive and simplified account. Yet during the period between Edward I's Expulsion of the Jews in 1290 and the 1655 Whitehall Conference under Oliver Cromwell only two types of Jews are recognisable in England: secret Jews and the imaginary Jews invented by the country that had banished them. The apparent absence of an actual Jewry only added to the proliferation of mythical accounts characterising Jews as villainous, and such conspicuous absence could not staunch the exaggeration and elaboration of this link with blood. Indeed, by the third century after the Expulsion the myriad 'blood-stories' had fused and developed into one lengthy narrative in which all evil acts perceived as characteristically Jewish (usually acts of spilling the blood of others or shedding it themselves) could be traced back to a discrete incident in the Bible (especially Judas' selling of Christ's blood and the Jews' behaviour at the Crucifixion) that neatly explains and augments the image of the demonic Jew. Whether medieval authors intended to strengthen popular scorn for the expelled Jews or simply employed extant hatred for material may be difficult to distinguish. Jews in English literature during this time were uniformly subject to depictions as demons, thieves, murderers, and perverts; there is little material before the millenarianist movement of the mid-seventeenth century contrary to this trend.[1] Emerging from this body of literature is a fictional figure that proves as dangerous to Christians as he is different, and representations of the Jewish apparition typically utilised this symbolic relationship to blood as a means of firmly marking those distinctions. Apart from an association with usury, the Jew in the medieval English mind was branded with allegations of ritual child murder and host desecration. The practice of circumcision established Jews as physically distinct from Christian men, and contributed to their status as a concern for anatomists and physicians, for whom the classification of Jews as physiologically different from Christians served to empirically support a burgeoning national identity.[2] Even in this last case, blood is central; traditionally, imagined Jews were prone to haemorrhoidal bleeding, nosebleeds, and other afflictions of blood-flux that contributed to their dissimilarity. Blood defined the medieval English Jew.

By the seventeenth century many individuals did indeed believe that Jewish men experienced menstruation. In his 'Large Diatriba of the Jew's Estate', Thomas Calvert, a minister from York, in 1648 wrote that

> [t]here is an excellent relation, if it can be proved to bear its weight with truth; to shew the originall of Child-Crucifying among the Jews. *Cantipratanus* saith, he once heard a very learned Jew, that in his time was converted to the faith, say, that a certain Prophet of theirs when he was at point of death, did prophecie of the Jews thus: *Know ye* (saith he) *this for a most certain truth, that you can never bee healed of this shamefull punishment wherewith you are so vexed, but onely by Christian blood.* This punishment so shamefull they say is, that Jews, men, as well as females, are punished *cursu menstruo sanguinis*, with a very frequent Bloud-fluxe ... I leave it to the learned to judge and determine by writers or Travellers, whether this be true or no, either that they have a monthly Flux of Blood, or a continuall mal-odiferous breath (19–20, 31; emphasis in original).

Calvert's language points directly to menstruation. The 'frequent' flux of blood he describes is later clarified as a 'monthly' flux, and he further emphasises the menstrual imagery by insisting that not only women but both sexes experience the blood-flux. Here women's blood is extended to Jewish men, associating Jews not only with blood but also with femaleness. Calvert was writing just a few years before the invitation for Resettlement, and the anti-Semitic backlash against the *Humble Addresses*, Menasseh ben Israel's petition to Cromwell in 1655, took many forms. Indeed, the growing millenarianism in England during the seventeenth century and increasing religious pressure to convert readmitted Jews, exemplified by Calvert's 'Diatriba', provided the volatile environment in which such outlandish claims could be made. Blood-stories isolated the male Jew as sub-masculine, his threat to Christianity manifest in a perversion of feminine traits coupled with a decidedly masculine appetite for violence. Lacking integrity in faith, bodily control, and familial authority, Jewish masculinity embodies a set of weaknesses against which Christian patriarchal values are more clearly realised.

Host desecration

The anonymous 1378 miracle play *The Play of the Sacrament* dramatises the interweaving of the Jews' threat to Christianity with their peculiar link to blood as evidence of their potential menace. *The Play of the Sacrament* was written several centuries after the adoration of the host began to gain popularity in England, but the text clearly indicates an anxiety towards the vulnerability of the actual body of Christ in the hands of Jews. In the play, a group of Jews cunningly obtain a consecrated wafer and proceed to, among other things, stab it with knives and nail it to a post in a gleeful re-enactment of the Crucifixion. The characters then cast the mutilated host into an oven, and from the flames emerges a bleeding Jesus to whom they all repent. The scene in which the men torment the host is remarkable in its conflation of New Testament Jews and an imagined medieval Jewry; they literally become the crucifiers of Christ, visualising themselves inflicting the five wounds – 'and so shall we smyte theron woundys five' (378) – by abusing the wafer and nailing it to a cross. As they slice the host, however, it begins to bleed, and at the sight of the blood the Jews are stricken with terror: 'Ah! Owt! Owt! Harrow! What deuyll ys thys?/Of thys wyrk I am on were!/Yt bledyth as yt were woode, i-wys!' (401–3). The Jews in the play, having so delighted in the destruction of the body of Jesus, are immediately petrified not only by the impossible life the wafer holds but also specifically by the action of bleeding 'as yt were woode [mad]'. The 'madness' of the bleeding host suggests the Jews' feminised hysteria and a particularly female madness associated with both menstrual and parturient blood.

The act of abusing the host characterises Jews as eager to defile a Christian symbol as the body of Jesus himself. In preparation for the mutilation, the Jews imagine that the host is indeed the body of Christ:

> Yff that thys be he that on Caluery was mad red,
> Onto my mynd, I shall kenne yow a conceyt good:
> Surely with ower daggars we shall ses on thys bredde,
> And so with clowtis we shall know yf he haue eny blood.
>
> (369–72)

This character, Jason, aims to prove 'yff that thys be he that on Caluery was mad red' – that is, Christ – by stabbing the 'bredde' and drawing blood. This action intensifies the demonic characterisation of the Jew as he intends to consciously abuse Jesus physically as well as symbolically. Furthermore, it is the appearance of blood that Jason recognises as proof of divine presence in the wafer and, in so doing, he predicts the miracle precisely. They are then horrified by the blood as confirmation not only of Jesus's presence but also of the profundity of their misdeeds. The blood in this story serves as an indicator that the Jews' integrity of faith is fragile; at the sight of the blood they lose all vicious motivations, and with bleeding wounds the image of Jesus asks them to repent and convert. As in several other medieval miracle plays, the Jews in *The Play of the Sacrament* are linked to blood as an emblem of their weakness in faith as well as their guilt.

Ritual child-murder

One logical extension of the re-crucifixion of Christ through abuse of the consecrated host is the actual execution of a good Christian, uniformly in medieval accusations a male child. The infamous slaying of Little St. Hugh of Lincoln – which generated, among other accounts, Chaucer's *Prioress's Tale* – resulted in the annihilation of the Lincoln Jewish quarter in 1255. Holinshed later relates this event as part of a series of crucifixions in his *Chronicles*: 'For they vsed yearlie (if they could come by their preie) to crucify one christian child or other' (ii. 437). This yearly sacrifice usually coincided with Passover, of which the near concurrency with Easter threw into relief the alien (and by extension diabolical) ceremonies imagined by Christians. The synchronised celebrations of Easter and Passover sparked a great deal of speculation, and the impetus behind such accusations was quite similar to that behind claims of host desecration – that is, to represent Jews as desiring to re-crucify Christ as a sign of evil. Furthermore, the symbolic symmetry established by such accusations between the spring holidays of Jews and Christians was essential to grasping the Jew's perverse relationship to the divine: while Christians celebrated the rebirth of the Son of God, Jews ritually murdered the son of a Christian.

Stories of this sort of attack on young Christians by Jewish men seem to have been disturbingly popular. Indeed, Chaucer's *Prioress's Tale* stands out as perhaps the most well known iteration of a common tale aimed at frightening young children. 'The Jew' in most ritual murder legends exercises inexplicable hostility and random violence over even the most pious of little boys. Yet upon closer analysis certain themes associated with the murders emerge in a more nuanced figuration of the Jewish assailant. *The Prioress's Tale* offers a somewhat simplified narrative of this dreadful accusation. In it, a pious widow's son obsessed with the *Alma Redemptoris Mater* hymn infuriates a group of Jews who capture him, slit his throat, and throw his body down a privy to escape the hymn still miraculously issuing from his corpse. The tale is part of a larger genre of miracle stories in which the Virgin Mary is responsible for the discovery of the little martyr by allowing his lifeless body to continue the hymn. The Jewish threat is here unique in its transference of culpability (an unspecified number of Jews hire a murderer to actually kill the child) but typically brutal and perverse in its violation of the boy:

> Fro thennes forth the Jewes han conspired
> This innocent out of this world to chace:
> An homicide therto han they hyred,
> That in an aley hadd a privee place;
> And as the child gan forby for to pace,
> This cursed Jew him hente and heeld him faste,
> And kitte his throte, and in a pit him caste.

> I seye that in a wardrobe they him threwe
> Where as these Jewes purgen hir entraille.
> O cursed folk of Herodes al newe,
> What may youre yvel entente yow availle?
> Mordre wol out, certain, it wol nat faille,
> And namely there th'onour of god shal sprede,
> The blood out cryeth on your cursed dede (131–44).

The murder itself is quickly done by a faceless figure, but the Jews' real offence is their disposal of the body in a 'wardrobe' or privy '[w]here as these Jewes purgen hir entraille'. It is from this foul place that the martyr's voice ascends, the horror of his slaying augmented by the fetid place where his pure body remains.

The latrine in which Chaucer's martyr is found is as elemental to the tale as the Virgin's perpetuation of the song. In a Latin manuscript dated about fifty years after Chaucer's death the same miracle is retold, including a similar fascination with the ritualised killing and the victim's disgusting temporary grave:

> They thought they were obeying God, but in fact they were making a sacrifice not to God but to the devils of hell. It is usual for malice to cease after death, but although they had killed the boy, their malice did not come to an end: they threw his corpse into a place of the coarsest filth, where nature purges itself in secret (Kolve 420).

The privy is repeatedly linked to this particular Jewish crime with disturbing implications, reinforcing an association of Jews with a space of contamination and filth. Compounding the sin of killing a small boy is the Jews' purported habit of a distressingly scatological form of disposal. This persistent detail identifies Chaucer's Jews as quite distinct from the disembodied Satan who 'swells' into their hearts to inspire homicidal action:

> O Hebraik peple, allas!
> Is this to yow a thing that is honest,
> That swich a boy shal walken as him lest
> In youre despyt, and singe of swich sentence,
> Which is agayn oure lawes reverence? (126–30).

Satan goads them to fatally punish the boy for singing of a 'sentence' (subject) that goes against their shared 'lawes reverence', but the Jews further their physical embodiment of malevolence by choosing the basest of human spaces to hide their crime. The innocent blood of the victim is engulfed in the excrement of the assassins as a disturbing reminder of their bodily perverseness.

The blood-flux

The twin notions that the Jew delighted in, and that the Jewish faith required, the spilling of Christian blood certainly contributed to the vilification of the imagined Jewry. Yet these tales create a slightly subtler characterisation of the Jew as cursed by blood as much as capable of spilling it. The biblical origins of the myth (and indeed the justificatory proof-text for later accusations) are found in Matthew; when Pilate declares that he is innocent of the blood of Jesus, the Jews reply, 'His blood be on us, and on our children' (Matthew 27:25). This moment in the narrative led to a number of claims in medieval England, not the least of which being that Jews themselves bled routinely as punishment for their role in the Crucifixion. Contemporary Jews, to the medieval English subject, were just as guilty of the killing of Christ as their New Testament counterparts. As illustrated above in Calvert's seventeenth-century explanation of this phenomenon, by the early modern period English anti-Semites had little difficulty suggesting that all Jews menstruated, at least in the sense that 'Jews, men, as well as females, are punished *cursu menstruo sanguinis*, with a very frequent Bloud-fluxe.' The theological treatise *Jacob's Well*, written between 1440 and 1450, provides an example of the more prevalent medieval hypothesis on bleeding Jewish men; they bled *yearly* on Good Friday, as an indication of their guilt in the killing of Christ.

The author of *Jacob's Well* is clear about connecting the Jews' response to Pilate with contemporary physiological ramifications. In the story of 'The Canon and the Jew's Daughter,' a young canon falls madly in love with a Jewess who lives with her exceptionally strict parents. The two lovers decide a tryst may take place on Good Friday, when Jews are universally debilitated:

> þat here loue & sche myyt noyt come to-gydere but it were on good fryy-nyyt; for, þanne, alle iews bledyn benethyn for wrech of cristes deth. For, whan pylate seyde to þe iewys, 'I am vngylty of þe blood of crist,' þe iewys seydin ayen to pylate, 'his blood be on vs & on alle oure children!' þerfore, on good fryy-nyyt, alle þe iewys lyen & bledyn be-nethyn. & þat nyyt þe chanoun lay be þe iewys dowter.
>
> [That her love and she might not come together but on Good Friday night, for, then, all Jews bleed beneath for wretch of Christ's death. For, when Pilate said to the Jews, 'I am not guilty of the blood of Christ,' the Jews said again to Pilate, 'his blood be on us and on all our children!' Therefore, on Good Friday night, all the Jews lie and bleed beneath, and that night the canon lay with the Jew's daughter.] (Brandeis ch. 26, 16–22).

The author compresses a number of ideas into this brief passage that call for further explanation. The story twice mentions that 'alle iews bledyn benethyn' on Good Friday, a condition that inexplicably precludes the

Jewish father from keeping a close eye on his daughter. The statement that all Jews 'bledyn benethyn' hints at a menstruation-like genital bleeding, and the explanation for this blood is 'for wrech of cristes deth'. The author elaborates on this concise rationalisation by invoking the passage from Matthew, concluding that 'þerfore, on good fryy-nyyt, alle þe iewys lyen & bledyn be-nethyn'. *Jacob's Well* clearly accepts as premise of this story the notion that Jews suffer a unique affliction (*not* the menstrual flow that Jewish and Christian women share) that comes once a year on the anniversary of the Crucifixion. On this night the smitten canon chooses to consummate his love for the Jewess.

The story as a whole, however, raises some perplexing issues. Why would the Jewish father relax the watch over his daughter on this night? How does the blood-flux of an entire community provide the canon with increased access to the daughter? The tone in which the author describes the following morning suggests that this affliction is debilitating to the extent of rendering immobile the young woman's father, at least during the night: 'On þe satyrday, be þe morwen, þe fadyr rose be-tymes, whan his blood was staunchyd, & say þe chanoun in his dowterys bed with here' [On the Saturday, in the morning, the father rose betimes, when his blood was staunched, and saw the canon in his daughter's bed with her] (22–4). The daughter's bleeding, on the other hand, is only mentioned implicitly (in that 'alle iewys bledyn') and that whatever affliction she may be experiencing neither stops the canon from wooing her nor does it preclude her from sexual activity. While the father is debilitated until his blood is 'staunchyd', the daughter enjoys what is assumed to be a consensual tryst. The tale concludes when the Jews go to the church *en masse* to accuse the canon of his transgressions and are miraculously mute after he repents in his heart. The young woman, in a convenient coda, converts and becomes a nun.

The distinction between the male and female experience of this yearly blood-flux is subtle but bears great consequence. The father's bleeding robs him of his primary descriptive characteristic: his strictness and vigilance over his daughter. The daughter, however, is uninhibited by the yearly affliction. I see evidence here of a deliberate distinction between Jewish masculinity and femininity. The young woman, arguably familiarised with a regular blood-flux, is less hindered by this Jewish affliction than her male counterparts. The father, less accustomed to the experience of the blood-flux, loses control of his family during Good Friday and ultimately loses his daughter to Christianity. Jewish masculinity in this example is decidedly hampered by the curse of the Jews' role in the Crucifixion; the blood this father sheds as punishment for Jesus' death initiates his loss of familial control. The male Jew is not necessarily feminised here – especially given that the female character is afforded increased freedom as her father is incapacitated – but he is emasculated. His authority is clearly jeopardised by the 'night of menses', and the blood itself marks an alien condition that forces him to relinquish control over his family.

'One jot of blood': Shylock and the Menstruation Myth

The connections between *Jacob's Well* and Shylock and Jessica are obvious, but the text can also help illuminate the tradition that informs Shylock's downfall, outside the home, in *The Merchant of Venice's* trial scene. Shylock replicates the 'Jew' of *Jacob's Well* in his inability to prevent his daughter's sexual transgression. As Shylock's ability to exert patriarchal control over his family wanes, Jessica's liberation grows. Indeed, she leaves the house this night 'in the lovely garnish of a boy' (2.6.47), augmenting her ability to transgress both her gender and her faith by cross-dressing and then by marrying into Christianity. Like Abigail from Christopher Marlowe's *The Jew of Malta* and the Jew's daughter in *Jacob's Well*, Jewish women are available for both sexual desire and religious conversion. The mercurial Jewish daughter evades her strict father, whose authority is permanently constrained by his susceptibility to a conflicted gender.

Although Shylock is not depicted as experiencing a blood-flux the night he goes to Bassanio's feast, his predicament mirrors the Jewish father in *Jacob's Well* in his replication of the consequences associated with the 'night of menses', the loss of control over his family. Shylock's presence at Bassanio's feast is absurd, as he has already promised 'I will buy with you, sell with you, talk with you, walk with you, and so following, but I will not eat with you, drink with you, nor pray with you' (1.3.29–32). That Shylock cannot eat at a feast according to his faith parallels his inability to act as a father and maintain control over his family. In *Jacob's Well*, the 'night of menses' marks a key moment of impotency as a father, and Shylock's attendance at a feast that cannot feed him marks his own impotency as a patriarch. Furthermore, the mocking words of his former servant Launcelot places the time of the play around Good Friday. When Shylock recalls a portentous dream, Launcelot jibes him, saying 'I will not say you shall see a masque, but if you do, then it was not for nothing that my nose fell a-bleeding on Black Monday last at six o'clock in the morning' (2.5.23–6). Launcelot's gibberish ridicules Shylock's superstition over a dream, but he does so in a way that touches not only on the time of year (Black Monday, or Easter Monday, falls just three days after Good Friday) but also hints at a kind of blood-flux that links Shylock back to the Jewish blood-stories: the nosebleed. This is not to say that Launcelot is asserting these connections of time and affliction. Rather, this fool is decisively mocking them. He anticipates Shylock's fears of replicating the myth of the bleeding Jewish father and then dismantles the myth to deride Shylock, his former master. In so doing, however, he indeed predicts the actual events of the night. By suggesting this myth, he sets up Jessica's evasion of her father and her faith. Jessica's disappearance evokes the legend of a Jewish blood-flux, and she herself even asserts that 'though I am a daughter to his blood,/I am not to his manners' (2.3.18–19).

This circumstantial connection between the myth of the menstruating Jewish man and Shylock expands into the trial scene, where this connection is cemented. Shylock's desire to take his pound of flesh is deflated when Portia asserts the condition of a bloodless exchange.

> This bond doth give thee here no jot of blood;
> The words expressly are 'a pound of flesh.'
> Take then thy bond, take thou thy pound of flesh;
> But in the cutting if thou dost shed
> One drop of Christian blood, thy lands and goods
> Are by the laws of Venice confiscate
> Unto the state of Venice (4.1.304–11).

Here again Shakespeare both invokes and reverses the myth of the menstruating Jewish man. The very idea of blood destroys Shylock, and appears here as the mark of a transgression. *If* he spills any blood, he will go against the letter of his own bond. Because he cannot retract the bond, blood in the trial context makes it impossible for Shylock to participate in the masculine systems of Venetian law and economics. Shakespeare does not, however, apply this formula directly. The key reversal here is found in the stipulation of 'One drop of Christian blood'. Shylock's own blood is irrelevant; the blood of the Christian, however, becomes precious and seals his doom. Again, the resonance of Matthew's 'His blood be on us, and on our children' infuses Shylock's character with remnants of these Jewish stereotypes. Furthermore, like the entire Jewish community in *Jacob's Well*, Shylock's silence in the face of Christian 'justice' reaffirms his othered status. His mute exit makes painfully clear that – as a man and a Jew – he cannot be a part of the Christian Venice.

The myth of Jewish male menstruation, especially in its focus on material blood in physiological, religious, and criminal contexts, shows that menstruation in early modern England held a unique role as the embodied intersection of the components of identity formation. Although competing explanations of menstruation and menstrual blood's function in reproduction abounded, the period of menstruation proved, to the early modern mind, the inferiority of women. Menstrual blood was a kind of odorous excrement, and it served as evidence that women's bodies (and thus women) were leaky, unstable, corrupted, and above all weak. The myth of Jewish male menstruation breaks open a discourse on menstruation in which the negative connotations of menstruation were transformed and intensified when attributed to men. Men who menstruated, then, were feminised and doubly vilified according to their transgression of gendered expectations. What Shapiro's and Katz's important studies have overlooked, however, is the way that the myth of Jewish male menstruation demarcates an entire field of conflict for Shylock in the domestic as well as the public sphere. The myth of the menstruating Jewish man narrates the destruction of Shylock's family, wealth, and faith. Within the conflict between Christian England and a largely imagined Jewry, the prominence of menstruation reveals an English masculinity that insists upon differentiating itself from the narrative shared by Shylock and the Jew in *Jacob's Well*. As suggested by the evolution of Jewish blood stories from the violence of *The Prioress's Tale* to the humiliation of the mute Jewry of *Jacob's Well*, the resulting myth of Jewish male menstruation serves to belittle and demean the Jew more than to elicit fear and hatred. In this context, menstruation may be seen not as a fixed stain of femaleness and inferiority, but as a discourse in which gender, religion, class, and community are manipulated as fluid markers of difference.

Notes

1. With the Reformation came an escalating sense that England itself was the locus of God's attention and judgment and in fact the site of the Second Coming of Christ. The millenarianist movement included genuine aims to convert Jews to prove the purity of the Church and its preparedness for the messiah.
2. For more on the physiology of sexual difference and its relationship to menstruation during this period, see Stolberg, this collection.

Works cited

Anon. *The Play of the Sacrament. Chief Pre-Shakespearean Dramas.* Ed. John Quincy Adams. Cambridge: Riverside, 1924. 243–62.

Anon. 'The Story of the *Alma Redemptoris Mater*'. Trans. A.C. Rigg. *The Canterbury Tales: Authoritative Text, Sources, and Backgrounds, Criticism.* Ed. V.A. Kolve and Glending Olson. New York: W.W. Norton, 1989. 418–23.

Brandeis, Arthur, ed. *Jacob's Well: An English Treatise on the Cleansing of Man's Conscience.* London: Oxford University Press, 1900.

Bynum, Caroline Walker. *Holy Feast and Holy Fast: The Religious Significance of Food to Medieval Women.* Berkeley: California University Press, 1988.

Calvert, Thomas. *The Blessed Jew of Marocco: OR, A Blackmoor made White.* York: 1648.

Chaucer, Geoffery. 'The Prioress's Tale'. *The Canterbury Tales: Authoritative Text, Sources, and Backgrounds, Criticism.* Ed. V.A. Kolve and Glending Olson. New York: Norton, 1989. 207–14.

Holinshead, Raphael. *Chronicles of England, Scotland, and Ireland.* 1586. London: J. Johnson, 1807–8.

Katz, David S. *Philo-semitism and the readmission of the Jews to England, 1603–1655.* Oxford: Clarendon, 1982.

——. 'Shylock's Gender: Jewish Male Menstruation in Early Modern England'. *Review of English Studies: A Quarterly Journal of English Literature and the English Language* 50.200 (November 1999): 440–62.

Marlowe, Christopher. 'The Jew of Malta'. *Doctor Faustus and Other Plays.* Ed. David Bevington and Eric Rasmussen. Oxford: Oxford University Press, 1998. 247–322.

Shakespeare, William. *Macbeth. The Norton Shakespeare.* Ed. Stephen Greenblatt. New York: Norton, 1997. 2564–617.

——. *The Comical History of the Merchant of Venice. The Norton Shakespeare.* Ed. Stephen Greenblatt. New York: Norton, 1997. 1081–146.

Shapiro, James. *Shakespeare and the Jews.* New York: Columbia University Press, 1996.

The Holy Bible in the King James Version. Nashville: Thomas Nelson, 1984.

16

A Menstrual Lesson for Girls: Maria Edgeworth's 'The Purple Jar'

Hollis Robbins

Maria Edgeworth's celebrated short story, 'The Purple Jar' (1796), concerns a young girl who becomes entranced by a purple jar she sees in a shop window. Of the dozens of stories Edgeworth wrote in the late eighteenth and early nineteenth centuries, this parable of desire and disappointment has retained the strongest hold on the public and literary imagination and was one of the most widely read, anthologised, and debated children's tales well into the twentieth century. It is considered by children's literature scholars to be a cultural touchstone, an epitome of the sort of moral and didactic children's tale that flourished until around the time of the First World War. However, the reason for the story's initial appeal has not yet been a subject of sustained investigation. Most scholars simply remark on its emblematic status and its tendency to anger female readers. A close examination of 'The Purple Jar' and its reception by female readers reveals that women recognized and responded to the story's coded subject matter – menstruation – and, specifically, the manner by which girls learned of their menstrual 'fate'.

In the story, a young girl, Rosamond, wants a purple jar so much that she chooses it over a new pair of shoes. 'You might be disappointed,' her mother cautions, adding that she will not be able to afford new shoes until the next month (142). Rosamond persists nonetheless: 'You can't think how these hurt me; I believe I'd better have the new shoes – but yet, that purple flower-pot' (143). When the purple jar is delivered to her home later that day, Rosamond examines it closely and discovers that it is merely a clear glass jar filled with blackish-purple, disagreeable-smelling liquid. 'Oh, dear mother!', she says, when she opens the jar. 'But there's something dark in it; it smells very disagreeably; what is it?

213

I didn't want this black stuff' (144). Rosamond's mother responds to her tears with indifference and some anger, explaining that 'the best thing is, to bear your disappointment with good humour' (144). As foretold, Rosamond is unable to play, dance, or even walk comfortably in her ragged shoes. The tale ends with her father refusing to take her out in public because she looks slovenly. 'Oh, mamma,' she blushes, 'how I wish that I had chosen the shoes! They would have been of so much more use to me than that jar: however, I am sure – no, not quite sure – but I hope I shall be wiser another time' (145).

The numerous allusions to and critical analyses of 'The Purple Jar' have emphasised utility and morality. A 1902 article on soda-drinking by children in *Medical News* alluded to the story in saying that children want what they want whether it is bad for them or not: '[t]here were few tucks to let down in the hem of Rosamond's garments when she realized the inutility of her "purple jar" – so we fear that the warning of the Chicago chemist will avail but little … for youth will have its fling, and its soda too' (Anon., 'The Proper Drinking' 1190). Another strand of 'The Purple Jar' criticism registers the story's preoccupation with social morality, public places, and the marital marketplace. Bernice E. Leary introduces her discussion of Edgeworth's work by suggesting that the ending of the story shows 'the natural consequences of humiliation and slovenliness' (par. 25). A 1939 journal entry by actress Fanny Kemble also situated Rosamond in the context of cleanliness and public restrooms:

> Now I counted no less than seven handsome looking glasses on board of this steamboat, where one towel was considered all that was requisite, not even for each individual, but for each washing room. … It is the necessary result of a young civilization, and reminds me a little of Rosamond's purple jar, or Sir Joshua Reynolds's charming picture of the naked child, with a court cap full of flowers and feathers stuck on her head. (6–7)

Elizabeth Gaskell's *Mary Barton* (1849) alludes to Edgeworth's story in the context of enchanted gardens, wild romances, temptations, abandoned girls, and cold, flowing rivers. Australian poet Ursala Bethell's poem 'By the River Ashley' also refers to 'flows' in Edgeworth's 'The Purple Jar'. Bethell's 'water of illusion' refers directly to the coloured water that Rosamond pours out of the jar. Heidi Thomson also suggests that Bethell situates the story within the context of reproduction: 'Like Ursula Bethell, Maria Edgeworth was closely involved with the education of children, mostly her own twenty-one siblings (her father married four times)' (47). Roderick McGillis has argued that Edgeworth's story is a capitalist tale, designed to socialise young readers into modern market practice: '[w]e can see that a story such as Maria Edgeworth's *The Purple Jar* takes pains to warn the child reader of the temptations of the market' (67). This range of opinions indicates the strength of opinion surrounding 'The Purple Jar' and provides an account of the cultural after-life of this text. For the most part, its reception has been filled with implicit acknowledgments of its coded subject matter. These can help delineate the meanings of menstruation for Edgeworth.

Rosamond's name, the appearance of the sticky black liquid, and the story's emphasis on monthly spending all establish the menstrual theme. The items that Rosamond desires, 'roses, and boxes, and buckles, and purple flower-pots', are terms with well-documented menstrual, sexual, and procreative connotations. Moreover, the fact that her mother encourages her to choose shoes instead suggests that she cannot escape what awaits her: 'shoes' was an alternative eighteenth-century spelling for 'shows' or menstrual blood. Rosamond's parents emphasise her reproductive future, using terms such as 'use' and 'useful' ('use' as a verb being euphemistic for 'to have sexual intercourse with', and as a noun for 'employment or maintenance for sexual purposes'), 'old shoes' (thrown at newly married couples for fertility and good luck since the fifteenth century), 'soles' (celibate men or unmarried females), and 'slip-shod' (slatternly, wearing bedroom slippers in public, or even 'undressed').[1] Rosamond is going to menstruate regardless of her choices. The text's problem is that she wanted to learn it as a precocious young girl rather than late in her teens, which would be more properly neat, modest, and well-bred. The painful lesson for Rosamond – the choice between shoes or flowers – is less concerned with making bad choices than with causing her mother to be angry and embarrassed. Rosamond's sorrow and the tale's poignancy is the loss of her romantic vision and her realisation that growing up is about usefulness, not beauty. As part of a narrative dealing with the idea that growing up is concerned not with beauty but usefulness, before she sees the purple jar, Rosamond wants 'beautiful roses' (141), and is downcast when her mother does not want them as well; however, by the end of the story, she has been utterly disenchanted. The story is clear in its lesson that a young lady should delay learning about 'disagreeable' necessities, should regulate herself, should keep herself from being surprised, and should not appear slovenly before the male members of her family.

In the nineteenth century, this is precisely what books such as F.L.'s *The Female Friend* (1809) advised in their, albeit rare, mentions of menstruation. Daughters were taught to be neat and clean. Moreover, it was made clear that they should not talk about sexual matters, including menstruation. However, *when* and *how* they should be taught about menstruation was rarely mentioned. Most 'good housewife' texts circulating in eighteenth and early nineteenth-century Britain and America – including William Buchan's *Domestic Medicine* (1769), Daniel Defoe's *The Family Instructor* (1715), William Alcott's *The Young Wife; or, Duties of Woman in the Marriage Relation* (1837) and Lydia Maria Child's *Mother's Book* (1831) – were silent on the subject of menstruation. Even texts that addressed the subject of menstruation, such as 'Aristotle's' *The Midwife's Guide* (1684), A.M. Mauriceau's *The Married Woman's Private Medical Companion* (1845) and Henry Arthur Allbutt's *The Wife's Handbook* (1896), were silent on how mothers should broach the subject with their daughters. John Dye's *Healthy Mothers and Healthy Children* (1904) cautioned that a proper mother 'should be solicitous for the welfare of her daughter' at the critical period of her development into womanhood, 'for carelessness then may cost her life. She should be instructed what she may expect at such a time, lest suddenly surprised she may do something to her disadvantage' (251–2). Just *how* she should be instructed is left to the mother.

Moral lessons and public education

Whether or not nineteenth-century readers recognised the story's implicit subject-matter, 'The Purple Jar' immediately made its way into the hands of educators in Britain and the United States. Austin Dobson reported that contemporary reviews were 'highly laudatory'. For him, Edgeworth's stories

> are happily contrived to excite curiosity and awaken feeling without the aid of improbable fiction or extravagant adventure. The language is varied in its degree of simplicity, to suit the pieces to different ages, but is throughout neat and correct; and, without the least approach towards vulgarity or meanness, it is adapted with peculiar felicity to the understandings of children. (51)

The young Princess Victoria was later assumed to have been reading Edgeworth's story in the 1820s:

> We can guess with tolerable certainty what was the Princess's child-world of books. ... [it] was still the age of Mrs. Barbauld and Miss Edgeworth.

'Evenings at Home,' 'Harry and Lucy,' and 'Frank and Rosamond,' were in every well-conducted school-room. All little girls read with prickings of tender consciences about the lady with the bent bonnet and the scar on her hand, and came under the fascination of the Purple Jar. (Tytler 210)

In 1836 an anonymous reviewer of *Rosamond; With other Tales* asked confidently, 'what child does not know Rosamond, – her purple jar – her knotted knight-cap – her little affectations?' (Anon. Review). 'The Purple Jar' turns up persistently in short magazine fiction about mothers without illusions – the only sort of mother capable of teaching her daughter the sad facts of life. A mother in an anonymous story, 'The Watch' (1834), offers her daughter a choice between a 'useful' doll that doubles as a pincushion, and a watch that does not keep time. After the painful lesson is learned, the mother says to the crestfallen girl, 'while you are sewing, I will read to you an account of 'little Rosamond and the purple jar' (143). A short story from *Godey's Lady's Book*, 'Mrs. Murden's Two-Dollar Silk' (1854), has a parallel between Rosamond's desire for the purple jar and a young wife's desire for a beautiful – but expensive – purple silk. When her husband is forced to give up a new overcoat to pay for the dress, the young woman chooses to walk to church with him every week in a shabby, plain old dress, 'a humbler and a better woman' (Anon. 'Mrs Murden's' 322).

Criticism of 'The Purple Jar' was largely concerned with two things: whether Rosamond's desire for the purple jar is indicative of her lack of virtue and whether her mother's reaction to her disappointment was appropriate. Home-schooling advocate Charlotte Mason hinted in 1886 that seven-year-old Rosamond is somehow naughty to have wanted the purple jar: '[l]ittle girls do not often pine for purple jars in chemists' windows; but that we should suffer for our wilfulness in getting what is unnecessary by going without what is necessary, is precisely one of the lessons of life we all have to learn, and therefore is the right sort of lesson to teach a child' (lx). An unsigned essay on 'Children and Modern Literature' in an 1892 issue of *The Living Age* declared that 'we are less angry with Rosamond for admiring the purple jar in the chemist's window than with her mother for permitting the child to buy it', suggesting that the mother should have been even more strict (290). Soon after the turn of the century, child advocates Kate Douglas Smith Wiggin and Nora Smith wrote that '[p]eople may say what they like of Miss Edgeworth's lack of proportion as a moralist and economist but we have few writers for children at present who possess the practical knowledge, mental vigor, and moral force which made her an imposing figure in juvenile literature for nearly a century' (147) adding that, because of the forceful figure of Rosamond's mother, no other stories 'have ever had a more remarkable influence upon young people' (148). Some woman writers reacted with anger to the lack of sympathy exhibited by Rosamond's mother. Rose, the heroine of Louisa May Alcott's *Eight Cousins* (1875), reacts indignantly to the suggestion that she should learn from experience like Rosamond and claims that she thought Rosamond's mother was unfair not to warn her daughter. A character in Edith Nesbit's *Wet Magic* (1913) similarly refers disparagingly to Rosamond's mother and Eudora Welty, in her short story 'A Sweet Devouring' (1957), indicts Rosamond's mother for not providing appropriate guidance:

> In The Purple Jar, it will be remembered, there was the little girl being taken through the shops by her mother and her downfall coming when she chooses to buy something beautiful instead of something necessary. The purple jar, when the shop sends it out, proves to have been purple only so long as it was filled with purple water, and her mother knew it all the time. (797)

The anger felt against Rosamond's mother may be partly ascribed to the tensions surrounding the taboo subject of menstruation. Mary Jane Lupton has noted that daughters' 'menstrual anxiety' can be partly ascribed to their mothers' 'secrecy or to her personal negligence' (162). The reaction to the mother in 'The Purple Jar' could be understood as a reaction against the way in which menstruation was a topic rarely discussed between mother and daughter.

Menstruation, morality and reproduction

Edgeworth's 'The Purple Jar' is positioned at moment of transformation in the moral and medical discursive history of menstruation, particularly in relation to the role of female education. This moment of transition is apparent in the nineteenth-century response to the story. In the eighteenth century, as Alexandra Lord (*passim*) and Richard C. Sha (par. 28–9) have pointed out, menstruation began to be cited as a key to understanding the female body and 'female complaints'. Doctors and male midwives increasingly discussed the subject among themselves, in medical journals, and with their female patients, collecting and sharing data. Menstruation, as a branch of obstetric science, preoccupied medical philosophers, anatomists, and physiologists. One doctrine that many physiologists agreed upon, according to James Manley, was 'that the menstrual discharge differs from blood, in being of a thicker consistency – in possessing a different odour – in having a much darker colour' (57). The nature of menstrual blood had moral implications: if menstrual blood differed from the blood that flowed through a woman's body (and in particular flowed to her cheeks), it was perhaps not to be considered an appropriate marker of virtue. Eighteenth-century doctrines of sensibility dictated that a (white) woman's colour represented her moral state. She should faint when shocked, be properly rosy-cheeked when healthy, and blush when embarrassed. If menstrual blood were the same as other blood, then it should likewise be affected by her moral nature. Thus, a 'moral' woman ought to have regular periods. Irregular menstruation betokened imprudent or unseemly thoughts, or perhaps indicated something amiss in the bedroom. A second, and perhaps more important, moral issue for medical philosophers was the apparent difference in the onset of menstruation in countries of different climates: women in hotter climes began their cycles early and those in northern – that is, 'civilised' – climes, began at a later age. Only 'savages' menstruated at a young age. Early menstruation was a sign of improper (that is, inappropriately early) sexual desire. The best cure for this was understood to be marriage and childbirth.

What, however, should be done to ensure that young women menstruated properly? Somebody ought to be watching. The best person to oversee that a young woman was menstruating properly – neither too early nor too often – was the mother. Yet, as mentioned above, mothers had little or no written guidance as to how to advise their daughters. We can assume that advice on how to educate daughters was passed informally, between female relatives and midwives. Teaching a daughter in this vacuum of 'approved' information would have made this exchange of information difficult and it possibly ensured that it was put off for as long as possible. However the information arrived, most daughters would have had a specific (anti-)learning moment, in which she learned not just menstrual etiquette but that the subject was not one to be learned about. The lesson of both 'The Purple Jar' and the responses of its female readers is that a young girl should not ask about it too early. Rosamond's mother tries to warn her off the subject: '[p]erhaps, if you were to see it nearer, if you were to examine it, you might be disappointed', she says, with a hint of malevolence as Rosamond examines the purple jar that has been delivered to her home (144). She meets Rosamond's disappointment with indifference. This reaction met with anger from some of the readers: 'I hate, I simply detest that mother of Rosamond', Edgeworth's biographer Emily Lawless complained in 1905. 'If at this climax of her sorrows poor Rosamund "retired and burst into tears", it is hardly to be wondered at; indeed, but for pure wrath, I suspect that the listener would have done so likewise!' (Lawless 49–50). But in an era when menstruation was not openly discussed, Edgeworth's depiction of the relationship between the mother and daughter would have been familiar to many women.

One of the perceived difficulties in educating young girls about menstruation was that the lesson was necessarily also about human reproduction. Girls had to learn that their wombs – jars or vessels – will someday be filled with a child. Even more difficult would be explaining how this occurs. Many of the nineteenth-century allusions to 'The Purple Jar' recognise the connection. An 1862 magazine story, 'The Plain Woman's Story', portrays a serious young man conversing with a plain woman:

> I am generally dubious about very bright colors. In twelve years, or so, the bright colors are usually washed out, and, like Rosamond's Purple Jar (which, I suppose, you remember?) there is nothing left but a common vessel. The man, too, who has been taken with this brightness of coloring, as Rosamond was taken with her jar, is very apt to think regretfully of some plainer woman whom he, perhaps, slighted in those early days when he did not think of acting for the future. (Rodman 407)

Here it is clearly the vessel which is important, not the purple substance contained within the vessel. This emphasis on the vessel is picked up by the British painter Henry Tonks's *Rosamond and the Purple Jar* (see Figure 16.1), exhibited at the Tate Gallery in London in 1900. The painting offers a startling image of a girl somewhat older than seven, although still pre-pubescent, in a dark green dress, arms raised, walking toward a large clear ornamental glass jar filled with a reddish liquid. While Edgeworth's story is of the eighteenth century, Tonks's work is clearly of the nineteenth. Unlike the small apothecary jars of Edgeworth's era, Tonks's glass decanter is large and Victorian: almost two feet high, delicate, and ornate. The liquid is a bright red rather than purple. The lid is wide and tall, the jar resembling that of a chemist or sweet shop window. Since Tonks's jar and the broad base are clear glass, it seems unlikely that anyone would mistake this jar of coloured liquid for a coloured jar. The jar is also shaped like a uterus: rounded, wider at the top than the bottom, with gravity pulling the liquid downward into a narrow tube between the jar and the table.

Figure 16.1 Henry Tonks, *Rosamund and the Purple Jar* (1900).

Conclusion

This short story articulates the complex relationships between a girl, her mother, her father and her duty to prepare herself for womanhood. It is, thus, not only an account of menstruation but also an examination of how the discourse of menstruation functions. Obliquely represented, menstruation here is positioned as resulting in sexual knowledge and sadness, as Rosamond finds she must take her place within the patriarchal hegemony. It is, after all, her father who is quick to punish her. It is his voice that crushes her at the end of the story: ' "[w]hy, are you walking slipshod? No one must walk slipshod with me! Why, Rosamond," said he, looking at her shoes with disgust, "I thought that you were always neat? Go, I cannot take you with me" ' (145). Maureen O'Connor has argued that in much of Edgeworth's work 'it is maternity rather than paternity that determines identity' (410). The severity of the mother in this story can be understood as a critique of the patriarchy, which resonates with Edgeworth's self-articulation:

> [She was] resistant to the aggressive, masculine norms of adulthood. ... not only did she model her most famous child character, Rosamond, on herself, but her insistence on regarding herself as 'little i' as a grown woman and an avowed preference for daughterhood also can be seen as claims to not simply childishness, but also an implied androgyny, qualities that allow evasion of the patriarchal imperatives of marriage and motherhood. (O'Connor 415)

Marriage, motherhood, maternal guidance and menstruation are recurrent tropes in this story, all contained by the parameters of the patriarchy. Rosamond's sorrow at discovering that the jar is not what she thought it was should be read as sorrow at comprehending how sexual difference functions as a mechanism for the cultural containment of female agency.

Note

1. Also, in the eighteenth century, 'to tread one's shoe awry' means to lapse from virtue. Reading along these lines, phrases such as 'Mr. Sole, the shoemaker', whose shop disgusts Rosamond, read differently.

Works cited

Alcott, Louisa May. *Eight cousins; or, The Aunt-Hill*. Boston: Little, Brown, 1917.
Alcott, William A. *The Young Wife; or, Duties of Woman in the Marriage Relation*. Boston: G.W. Light, 1837.
Allbutt, Henry Arthur. *The Wife's Handbook: How a Woman Should Order Herself During Pregnancy and After Delivery*. 50th ed. London: Forder, 1926.
Anon. 'Children and Modern Literature'. *The Living Age* 192.2483 (30 Jan. 1892).

Anon. 'Mrs. Murden's Two Dollar Silk'. *Godey's Lady's Book*. (Apr. 1854): 48.

Anon. 'The Proper Drinking of Adults'. *Medical News* 80.25 (21 June 1902): 1190.

Anon. Review. *Southern Rose*. (2 April 1836).

Anon. 'The Watch'. *The Juvenile Miscellany*. 6.2 (May/June 1834).

Aristotle [pseud]. *The Midwife's Guide*. New York, 1845.

Bethell, Ursula. *Collected Poems*. Ed. Vincent O'Sullivan. Wellington: Victoria University Press, 1997.

Buchan, William. *Domestic Medicine*. London, 1769.

Child, Lydia Maria Francis. *The Mother's Book*. Boston: Carter, Hendee & Babcock, 1831.

Defoe, Daniel. *The Family Instructor*. 1715. New York: Scholars' Facsimiles & Reprints, 1989.

Dobson, Austin. *De Libris. Prose & Verse*. New York: Macmillan, 1908.

Dye, John H. *Illustrated Edition of Painless Childbirth; or Health Mothers and Healthy Children*. 15th ed. Buffalo: Hausauer, Son & Jones, 1904.

Edgeworth, Maria. 'The Purple Jar'. 1796. *From Instruction to Delight: An Anthology of Children's Literature to 1850*. Ed. Patricia Demers and Gordon Moyles. Oxford: Oxford University Press, 1982. 141–5.

F.L. *The Female Friend; or The Duties of Christian Virgins. To Which is Added, Advice to a Young Married Lady*. Baltimore, 1809.

Gaskell, Elizabeth. *Mary Barton: A Tale of Manchester Life*. 1849. 3rd ed. London: J.M. Dent & Sons, 1961.

Kemble, Fanny. 'Journey from Philadelphia to Portsmouth VA. (21–22 Dec 1839)'. *Journal of a Residence on a Georgian Plantation in 1838–1839*. Athens: Georgia University Press, 1984. 12–13.

Lawless, Emily. *Maria Edgeworth*. New York: Macmillan, 1904.

Leary, Bernice E. 'Milestones in Children's Books'. *Books at Iowa* 12 (Apr. 1970). June 2004. <http://www.lib.uiowa.edu/spec-coll/Bai/leary.htm>.

Lord, Alexandra. ' "The Great Arcana of the Deity": Menstruation and Menstrual Disorders in Eighteenth-Century British Medical Thought'. *Bulletin of the History of Medicine* 73 (1999): 38–63.

Lupton, Mary Jane. *Menstruation & Psychoanalysis*. Chicago: Illinois University Press, 1993.

Manley, James R. 'Remarks on Menstruation'. *The New York Medical and Physical Journal* 4.1 (Jan.–Mar. 1825): 52–70.

Mason, Charlotte. *Home Education*. 1886. Oxford: Scrivener, 1955.

Mauriceau, A.M. *The Married Woman's Private Medical Companion*. New York, 1854.

McGillis Roderick. ' "Captain Underpants is My Hero": Things Have Changed – or Have They?' *ChLA Quarterly* 27 (2002): 62–70.

Nesbit, Edith. *Wet Magic*. 1913. New York: SeaStar, 2001.

O'Connor, Maureen. 'Maria Edgeworth's Fostering Art and the Fairy Tales of Oscar Wilde'. *Women's Studies* 31 (2002): 399–429.

Rodman, Ella. 'A Plain Woman's Story'. *Peterson's Magazine*. 41.5 (May 1862): 407.

Sha, Richard C. 'Scientific Forms of Sexual Knowledge in Romanticism'. *Romanticism On the Net* 23 (Aug. 2001) June 2004. <http://users.ox.ac.uk/~scat0385/23sha.html>.

Thomson, Heidi. 'A Note on Maria Edgeworth and Ursula Bethell'. *Kotare – New Zealand Notes & Queries* 2.1(May 1999): 46–7.

Tytler, Sarah [Henrietta Keddie]. *The Life of Her Most Gracious Majesty the Queen.* Ed. Ronald Gower. London: J.S. Virtue, 1885.

Welty, Eudora. 'The Great Devouring'. 1957. *Stories, Essays & Memoirs.* New York: Library of America, 1998. 797–803.

Wiggin, Kate Douglas, and Nora Smith. *Children's Rights: a Book of Nursery Logic.* Boston: Houghton Mifflin, 1892.

17
'A Rag and a Bone and a Hank of Hair': The Menstrual Background of 'the Vampire'

Andrew Shail

The 1915 Fox Film *A Fool There Was* initiated a film discourse figure known as 'the vampire', only later shortened to 'vamp'. This film was based on three sources: Philip Burne-Jones's 1897 painting 'The Vampire' (see Figure 17.1), his cousin Rudyard Kipling's poem 'The Vampire' (based on Burne-Jones's painting and included in the catalogue to its first exhibition) and Porter Emerson Browne's 1907 stage version of the poem, *A Fool There Was*.[1] 'The vampire', the stock bad woman character, was one of the most tenacious film archetypes of the rest of the silent period. Nita Naldi and Alla Nazimova were amongst those taking on the role habitually. The first of a series of vampire-role films for Theodosia Goodman (publicised as Theda Bara), *A Fool There Was* saw the most energy to date focused into the manufacture of a 'picture personality' persona. Rumours that the name was an anagram for 'Arab Death' were encouraged, and stories circulated concerning a mysterious origin in 'the Middle East' involving weaning on snake's blood, intensified by Bara often pretending not to speak English when appearing in public. If, as Marie Mulvey-Roberts has argued, the vampire of Bram Stoker's *Dracula*, in addition to echoing previous vampires by drawing on a history and pre-history of menstrual taboo, personified the morbid medicalisation of menstruation enacted by nineteenth-century discourse on the female body (82–3), this later vampire can be seen as a description of thinking on the menstrual body roughly twenty years after *Dracula*.

Burne-Jones, Kipling and Browne's female vampire has been attributed to the application of the discourse of industrial management to sexed identity, combined with more general fears concerning female

Figure 17.1   Philip Burne-Jones, *The Vampire* (1897).[2]

sexual license.[3] Constitutionally anathema to production, femaleness drains male production engines and production heads of their constitutionally productive energies, sucking out sperm and misappropriating it as a nutrient instead of a generative substance. A menstrually-weakened female constitution was just one component of this notion of femaleness. But if this account of sex provides a ready explanation for understandings of femaleness referenced by nineteenth-century gothic literature producing both male and female vampires,[4] changes in popular concepts of the sexed body, already taking place before Stoker's *Dracula*, had reached a point by 1915 which this version of the sexed body does not adequately explain. The connections between images of vampirism and thinking on menstruation are enduring,[5] but attention to what each vampire shows us about contemporary configurations of menstrual subjectivity is also important. The film vampire of 1915 certainly expressed beliefs about the menstrual body. Kipling's phrase 'a rag, a bone and a hank of hair' (220) circulated freely in popular film discourse, continuing to reference the menstrual morbidity of Stoker's Count Dracula, but the explanation behind this sexed morbidity was the result of relatively new widespread understandings of the body's material structure as well as the new representational mode of cinema.

Nerves

The nervous system was the basis of one strand of bodily re-discursification during the nineteenth and early twentieth centuries. With the specialisation 'neurologist' appearing in the 1830s, neurology became the most rapidly pursued branch of medical research in the nineteenth century. Its discoveries formed a major component of the widespread conversion from understanding organs in terms of mechanical functions that could be deduced in relation to the whole (the anatomical conception of the body) to investigating functions that organs might possess irrespective of their anatomical relation to each other (the physiological conception of the body). The transition to physiology followed from the first understanding of the cell – and therefore of the body as made of elementary living structural components – in the 1820s, a notion of functional ubiquity strengthened by the discovery of a body made up of different tissue types in the 1830s and 1840s with the emergence of histology.[6] The discovery that electricity was involved in nerve conduction in the 1840s was key to the solidification of the discipline of 'neurophysiology' in the 1850s,[7] and from this point medical language registered a turn to a conception of the nervous system as performing functions beyond the established case of muscle movement. This was reflected in a burgeoning of specialist terminology, from 'neuralgia' or nerve-pain in the 1820s to 'neurovascular', 'neurocrane' (describing the skull), 'neurorrhaphy' (nerve suture), 'neuropath' (someone with abnormal nervous sensibility), and 'neurotrophic' during the 1880s. The latter described the act of control of cells by nerves, particularly relating to the control and deficiency of cellular nutrition in explaining pathologies.

With the 'neuron' established as the discrete anatomical unit of the
nervous system and the reflex arc (where stimuli elicit responses without
conscious control) established as the basic functional unit of the nervous
system by 1906, nervous energy was tightly tied into notions of holistic
health. As John Chapman wrote in 1863, '[i]t is probable that the more
the diseases and functional derangements of animals having a nervous
system are investigated, the more they will be found to originate primarily in altered conditions of that system' (v). Michael Foster, the first
Professor of Physiology at Trinity College, Cambridge, hinged his 1885
definition of physiology on the nervous system. On receiving a sensation,
he wrote,

> sensory impulses reaching the central nervous system may forthwith
> issue as motor impulses leading to movement; but on many occasions
> they tarry within the central mass, sweeping backwards and forwards
> along particular areas of its substance, thus maintaining for a while
> a state of molecular agitation and leading to movement at some sub-
> sequent period only. ... Lastly, the presence of these molecular agita-
> tions in the central nervous system, whether the immediate result of
> some new afferent [sensory] impulse, or the much delayed and
> complicated outcome of some impulse which arrived long ago, or
> the product of internal changes apparently independent of all distur-
> bance from without and so far spontaneous, may be indicated by
> corresponding phases of what we speak of as consciousness. We are
> thus led to conceive of the central nervous system as, chiefly at least,
> the seat of a molecular turmoil. (9)

Foster's description of the body echoed the work of Thomas Laycock
from 1839 and that of W. Benjamin Carpenter from the 1840s. In the
1830s, reflex arcs in the spinal column, never troubling the supposedly
qualitatively different cerebral matter, were held by neurologists to be
exclusively responsible for all of the body's automatic functions.
Laycock, Carpenter and their respective followers drew attention to such
phenomena as the involuntary, in place of the voluntary, recollection of
memories, to indicate that the brain 'did not differ from the other ganglia of the nervous system' and that it preserved, from its past as just
part of the spinal column, a range of automatic functions in the form of
mental activities that were organised 'insensibly' (Laycock, 'On the
Reflex' 300). In addition to the functions of what is now called the 'autonomic nervous system' in the spinal column which could be guaranteed
never to reach consciousness, certain reflex actions were seen to occur

on the borders of consciousness, and to be affected by consciousness. The work of Laycock and Carpenter had significant influence on the discipline of neurophysiology, not least because it provided the basis for explanations of the brain as composed of the same cell types as the rest of the nervous system.

The discovery and diagnosis of neurasthenia from 1869 onwards was one consequence of the growing comprehension of this reflex function of the brain, and the term 'neurogenic', a corollary of the conception of the body 'as an assemblage of molecular thrills' (Foster 9), was first used at the end of the century to describe systems that were controlled by the nervous system. This was the culmination of a steady movement throughout the nineteenth century towards aligning most systems with neural control. When James Oliver explained menstruation as a nerve-origin phenomenon and not, as was being theorised, a shedding of the mucous membrane in 1887, he indicated not only the core role now played in the conception of the body by nerves, but also the lack of a distinction between conscious states and the nervous system being seen to control menstruation. He explained that '[a]ll visceral activities [activities of the vital organs] are now, through habituation, fulfilled in a somewhat automatic manner', adding that at previous stages of evolution they were connected with 'the higher cells of the cerebral lobes which participate in feeling' (379). Because the brain was just the evolutionary development of the spinal cord, '[l]ike all other nerve centres fulfilling a similar dispensation, this uterine centre is undoubtedly beyond all volitional control, but is nevertheless capable of being disordered by emotional impressions' (379). He thus linked menstrual disorder to 'a proneness to aggravation' (380).

The nerve-reflex theory of menstruation

By the 1850s it was 'widely accepted as one of the most incontestable facts of physiology' (Simpson 4) that menstruation (a) could not be understood as a secretion, and (b) that it had some connection to ovulation. Involving an explanation of menstrual blood as the unused nourishment of an un-fecundated ovum, this understanding led, in the words of one professor of medicine in 1870, to 'the recognition of the ovaries as the central, dominating organs of the whole female economy, whose general condition, both in health and disease, is modified in an infinite variety of ways by the changes, natural or morbid, which occur in them' (Simpson 4). By 1883 gynaecological opinion was moving towards a consensus on menstruation as the shedding of the mucous membrane of the uterus (as opposed to a rival contemporary theory of it as a 'freshening' of the uterus for the reception of an ovum) and so as 'a consequence of the non-occurrence of pregnancy' (Geddes 408). Concurrently the neurophysiological idea that the actions of the ovaries and the phenomenon of menstruation were actually consequences of another factor was growing in influence. This gradually lessened the recently-installed conception of the ovaries as the governing force in the female body.[8] Robert Lawson Tait, surgeon to the Birmingham Hospital for Women, was attributing menstrual irregularities to diseases of the nervous system as early as 1873. Edward John Tilt, author of five major works from the 1850s to the 1880s on menstruation, discussed menopause as the herald of ganglionic nerve disorders in his *The Change of Life in Health and Disease* (1857). John Chapman argued that his work with hot and cold shocks in the early 1860s 'constitutes a remarkable proof that the circulation of the blood in the womb is subject to the controlling influence of the sympathetic ["fight or flight"] nervous system; that the so-called functional diseases of that viscus are in reality abnormal conditions of the nervous ganglia which control it' (v). 'Dystrophies', defective nourishments of completely healthy parts of the body, were being related to a reflex action originating in nerves situated in damaged areas by the 1880s (Ord 4), and by 1887 James Oliver saw both epilepsy and 'interference with the outward manifestation of uterine activity' (380) (i.e. problems with menstruation) as the result of nervous disorder.

Hysteria had for some time been comfortably explained in neuro-physiological terms, where a menstrual constitution bordering on the pathological was seen to either drain women of nervous energy or to cause an imbalance of nervous energy. The increasing causal significance attributed to nerve-force, and the lessened distinction between conscious nervous state and 'autonomic' functions, led many practitioners to associate menstruation and hysteria only insofar as hysteria was related to neurotrophic disorder caused by the social anxieties attendant on menstruation. Uterine disorder was a consequence of this larger cause rather than a cause in itself. James Carruthers Young surveyed the most recent studies of hysteria in 1914 and concluded that the only satisfactory explanation for hysteria was that a hereditary predisposition to hyper-sensitive neurones was triggered by changes in the sympathetic nervous system brought on by mental stimuli (3–5). The potentially hysterical type was not sex-specific. Like the restoration of speech after the shock of a fire, hysteria, he argued, was 'purely psychic in origin' (25), a consequence of conscious trauma (14), including anything from worry to nervous breakdown (20), thus explaining its higher incidence in women, for whom menstruation was a further source of anxiety. Notions about nervous state had shifted so much that beliefs in women's constitutional nervous weakness or imbalance were being displaced by a conception of women as only subject to one more of hysteria's precipitating anxieties than men. T. Clifford Allbutt wrote of neurasthenia in 1910 that young women were much more prone than young men to neurasthenic breakdown (742). While continuing a coding-female of proneness to nervous disorder, Allbutt's comment indicates both the perceived proneness of the male body to nervous dystrophies and, by referring to *young* women, that the causes of neurasthenia also related not to a menstrual body *per se*, but to a nervous system encountering the social shocks of menarche.

Although the discourse of neurophysiology was not incompatible with the 'hysterization of women's bodies', discovering in the body marked female 'a pathology intrinsic to it' (Foucault 104), the idea that menstruation, problematic or not, was the *cause* of nervous disorder began to be exchanged for the notion that problematic menstruation was the *consequence* of nervous disorder. Whereas, in 1863, John Chapman's conclusion that the nervous system constituted the prime cause of functional derangements was led by a predetermining conviction of the instability of the female nervous system, by 1910 this was being modified. Menstruation's accompaniment by frequent pain meant that it was still being seen by 1883 and after to lie 'on the borderlines of pathological change' (Geddes 408), but the predominance of thought on nerves was pervading the classification of menstrual pathologies by 1885. William Ord concluded in this year that arthritis in women with excessive menstruation could not be substantiated by any notion of uterine secretion or septicity of the blood. He hypothesised that it was instead the result of 'nerve-agency', in this case 'a direct morbid influence of the spinal cord on the nutrition of the component parts' (4) of the body, where abnormal nutrition of the spinal cord, related not specifically to menstrual blood-loss but to increased flow of blood to the uterus, could, through 'reflex and trophic influence' (10), affect perfectly healthy parts of the body, including the joints, with such disorders as arthritis (20–1). Menstrual pain was explained within the same schema. Menstruation's previous morbidity was thus attenuated to pain understood as caused by 'centripetal nervous influence' (Ord 29) related only to anaemia. The existence of a 'menstrual wave', evidenced in variations in body temperature, pulse and blood pressure, was a subject of medical consensus by the end of the nineteenth century. Earlier theories of general bodily periodicity related to cycles of the sun, moon and seasons had been advanced as early as the 1840s, but they now waned in the face of nerve-reflex explanations. Consequently, as, even in the 1910s, the new science of endocrinology was only gradually formulating the most rudimentary elements of what would, after the war, begin to develop into the understanding of 'sex hormones' as the major influence over the menstrual and other bodily cycles, the commonly-accepted model, by the 1910s, of a rough 20–30-day rhythm (itself based on an only very recently identified norm of menstrual regularity) was still widely related to a nervous system not understood as sexed.

In addition, as Ornella Mosucci has shown, while the idea of a totally sexed ontology was the basis of much nineteenth-century medical thought on sex which gave rise to the discipline of gynaecology in the UK in the 1840s, many nineteenth-century medical practitioners and writers, far from widespread acceptance of dimorphic sex evolution, were just as careful to show, for example, that men and women possessed homologous structures for most supposedly sex-specific organs (16–20). After all, chromosomal and hormonal sexing were not part of their understanding of the body. William Stephenson, one of the respondents to Mary Putnam Jacobi's 1878 'On the Question of Rest for Women during Menstruation', disputed not her ideas of intrinsic menstrual health but the idea that the 'menstrual wave' was a matter of extremes, pointing out in 1882 that while menstruation was linked to a rise in body temperature, (a) menstruation was not the seat of this wave as the intermenstrual period saw a corresponding drop *below* the mean point in both body temperature and weight of urea excreted, and (b) menstruation itself occurred *after* the highest point reached by these variables, occurring not at the apex of either the positive or negative variation but around the mean point (Stephenson 9).

Thomas Laycock wrote in 1840 that

> the *phenomena* of menstruation ... consist in tumour of the mammæ, with a darker tint of the areolæ, weight and irritation about the pubes, pain in the loins, yawning, fastidious appetite, nausea, and not infrequently a sense of tension in the muscles of the neck, headache, and alternate pallor and redness of the cheeks: in addition, there is a flow of a sanguineous fluid from the vagina, varying in quantity from one to eight ounces. (*Treatise* 45, emphasis added)

While this total body conception of menstruation solidified with the steady discovery of the menstrual cycle after 1870, thus giving rise to lasting ideas about the female body as distinctly physiologically cyclical,

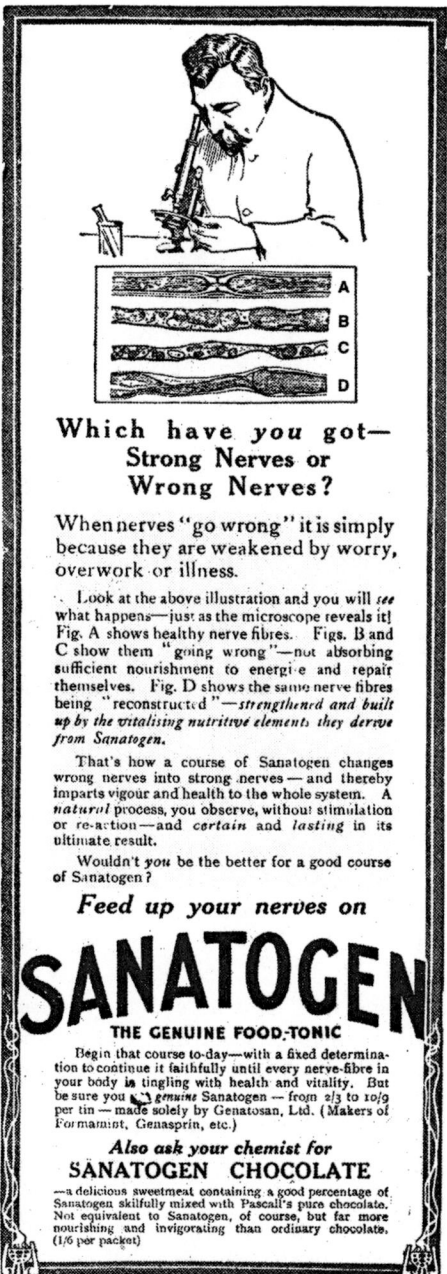

Figure 17.2 Advertisement for Sanatogen (October 1919).

periodicity as the consequence of nerve reflexes common to men and women simultaneously loosened this sexing of menstruation. For example, William Stephenson had also noted a corresponding 20–30-day wave in men, concluding that the phenomena being recorded as a wave 'belong not to the function of Menstruation but to a general law of Vital Energy' (10).

The properties of 'nervous energy', a term in common use by the first decade of the twentieth century, were not the subject of consensus. On one hand, the physiological conception of the body understood nervous energy as one of the forms the body engineered from the chemical energy in food. On the other, pathological conceptions of nervous energy tended towards the idea that it was an *excess* of nervous energy that was at the root of neuropathologies. Weakness was not equated with the dearth of a resource, and so conceptions of innate physical strength levels in male and female bodies did not translate to the condition of the nervous energy that was becoming increasingly central to holistic notions of bodily health, the 1910s witnessing an explosion of mass-advertised medication for 'bad nerves' and 'nerve-strain' (see Figure 17.2). By the middle of the nineteenth century the use of 'nervous' to mean 'muscular', 'vigorous', 'strong', 'energetic' or 'courageous' had receded with the onset of reference to agitation of the nerves. The intensified role played by mental state in physiological well-being, expressed in an explosion of lay terms pertaining to the central nervous system during the nineteenth century, was also witnessed by the transformation of 'nervous exhaustion', common in the mid-nineteenth century, into 'nervous breakdown' by the first decade of the twentieth century, a decade which also saw the emergence of 'nerve-wracking'. Both terms denoted a visceral situation rather than a change in state of mind, and both connoted understandings of the nervous system as far more instrumental in general health than when the expression 'to touch a nerve' was coined in the 1830s. By the first decade of the twentieth century, popular literature was peopled with characters whose states of nervous agitation were a part of the narrative. The time-specific diseases of the first two decades of the twentieth century were neuropathological, including 'spinal irritation', neuralgia and neurasthenia.

The film vampire

As a consequence of this sexed body, when a film journalist wrote in 1920 that the 'Vampire type' 'is comparatively new and very unhealthy, [and] has the graveyard odour' (Gore 372), this ill-health had been reconfigured. The vampire's menstrual subjectivity is implicitly witnessed in the first and last scenes of *A Fool There Was*. At the beginning, in a scene unrelated to the film's narrative, Bara's vampire character plucks the head off a flower and crushes it between her fingers, staining them with a liquid that, in monochrome, resembles blood. In the last shot of the film, the vampire sprinkles flower petals over 'The Fool' as he lies dying. 'Flowers' being one of the most common euphemisms for menstruation, these bracketing scenes bring her menstrual body to the fore. A previous notion of a morbid constitution was certainly instrumental in making the vampire a film type, as the fondness for Kipling's phrase indicated, but *A Fool There Was* and film discourse on the vampire indicated that this constitution now had a new basis in the discourse of neurophysiology. Bara acted the vampire as experiencing constant nervous agitation and frequent nervous fatigue, headaches and outbursts. Her influence on 'The Fool' perhaps best indicates her constitutional state. Perfectly capable of swinging a poker with strength even when about to die, he is suffering from nervous imbalance rather than drained life force or drunkenness.

Bara's invocation of the nervous body was certainly strong enough to signal her nervous disorder to reviewers. An article in the *New York Dramatic Mirror* in January 1915 was typical of descriptions of Bara's vampire, seeing her as 'a neurotic woman gone mad' ('Review' 4). Before Sigmund Freud's definition became widespread, 'neurotic' referred to predisposition to nervous imbalance. The nervous seat of Bara's vampirism was evidenced in the fact that, as one writer commented in 1917, '[y]ou are made to hear the tremendous wear and tear of her existence' (Codd 301). The fact that 'she is only a vampire under protest' (Codd 301) rooted vampirism in the body controlled by nerve reflex action. Vampirism had peeled away from a script of hostility towards others not just because it was serving as a myth of female sexual agency but because it was rooted in contagious nervous disorder rather than constitutional morbidity. In addition, the vampire's sexual availability, rather than simply a demonisation of female sexual agency, derived from what Austin Flint, describing nervous disorders in 1868, called 'a morbid susceptibility to emotions and a defective power of the will to restrain their manifestations' (694). One writer witnessed the menstrual root of vampirism when she announced in 1917 that 'we have come to the conclusion that no special qualifications are required for aspirants anxious to try the vampire stunt' (Codd 301), a universality recognised by vampire actress Nita Naldi in 1925 when she commented that 'every woman is potentially a vampire' (42).

Figure 17.3   Film acting as action heroism (*Pictures and the Picturegoer*, July 1914).

Cinema

The representational mode of cinema also 're-mediated' the menstrual body. As early as February 1913, an image of action heroism was being described in star publicity, one film journalist writing that Dorothy Foster 'does not know the meaning of "nerves", and calmly faces situations which would send most young ladies into a graceful faint' ('Miss' 13). 'I happen to be fairly strong and athletic,' Alma Taylor is quoted as telling an interviewer in 1914, 'so my rough-and-tumble adventures seldom worry me' ('The Girl ... 10' 443). Kathlyn Williams 'has run the gamut of Moving Picture sensations all the way from fly-ing machines to acting in a cage where there were untamed lions' ('Kathlyn' 31). This body configuration flowed from several sources. 'Action' was not a genre in the cinema of the 1910s. As a by-product of the turn to outdoor shooting in both the US and UK cinema industries in the 1911–14 period, it was threaded through films and film discourse, such as the above drawing from one popular British film fan magazine (see Figure 17.3), as a ratifying demonstration of film's non-parasitic and even superior relationship to the stage. Even stars whose film repertoire involved little action were the subject of litanies of near-death escapes in their publicity, an interview with British picture personality Elisabeth Risdon in October 1914 recording her astonishment that 'most cinema actresses ... seem to have suffered more hardships than would kill the average woman' ('The Girl ... 11' 152) before giving a litany of her own. As a consequence of the main impetus in film discourse from 1908 throughout the 1910s to move film away from its earlier classification as an amusement technology for making the familiar seem absurd or strange, film viewers were commonly moved during the 1910s to see the impression of perpetual movement attending the cinematic body – instead of attributing this sense of immortality to the transformative capacities of the apparatus – as deriving from its pre-filmic existence in the bodies of those being filmed. That nothing was faked means the conclusion was repeatedly drawn that 'the actors must have had a miraculous escape' ('Notes' 282). Consequently, even those members of the invincible cadre (beginning to be referred to as a constellation of 'stars' by 1915) whose selves, revealed in film, derived from the 'a rag, a bone and a hank of hair' statement of morbidity were the subject of this understanding of the body. Giving an account of Helen Holmes's 35-foot leap from a train into a river, one writer asked '[w]ho says a film-actress is nervous?' ('Picture News' 314).

Picture personalities were also widely seen to be playing themselves, one journalist commenting that '[t]he Alma Taylor of the screen is the

The "Cartonising" of Charlie.

The "Vampiring" of Mary.

VERSATILITY OVERDONE

Figure 17.4 As Mary Pickford's characters were all seen to be her angelic self, vampirism was a mystery to her (February 1918) ('News', *Pictures and the Picturegoer* 148).

Alma Taylor of real life' ('The Girl ... 10' 442), another that 'Mary Pickford is the same Mary on as off' ('Picture News' 38). Film articles would persistently refer to stars not as 'playing' but as 'being' vampires. (Naldi and La Marr 42) One writer quoted a friend of Bara's: 'The very first impression you get of Miss Bara is one of eeriness. She seems to glide into the room like a vapour' ('Vampires' 483). Vampire actresses were likewise seen to be playing themselves, and vampirism frequently understood as the preserve of specific temperaments (see Figure 17.4), one journalist referring to Theda Bara as 'that internationally famous vampire' ('Picture News' 2). Vampirism was seen to be an attribute not of the fictional body but of the filmed body, meaning that menstrual subjectivity was being forcibly understood via a cinematic apparatus unable to represent the body in terms of a sexed predisposition towards worn or imbalanced nerves.[9] Because of the innate health broadcast by the cinematic body, one writer, having described Bara's character with reference to 'her gliding, snake-like charm, her mysticism, her allure-ments', then went on to comment that also 'she has a sweet, child-like character' ('Vampires' 483). In addition, because cinema articulates its subjects as a plethora of surfaces and provides little direct concept of internality, its understanding of the body lessened the already consider-ably less brutal image of bodily disorder disseminated by nerve discourse. One writer commented in 1917 that '[n]erves are supposed to radiate from little grey bunches of things situated at either side of the spinal column and invisible to all. But the nerves now in fashion are not interior things. They are worn draped, swirled and slit ... all over' ('Here and There' 473). Nerves, it was apparent to this film journalist, were a pretence put onto a body that was essentially healthy, which charade she satirised by claiming 'you've got to have 'em, or you're horribly bourgeois ... don't you know' (473). Whereas Stoker's vampire reflected an understanding of the menstrual body as always in the process of dying, by 1915 this had dissolved. Now the verb 'to vamp', being used regularly by November 1917, expressed the menstrual subject, including a link between menstruation and will, and indicated a sense of the vampire being an action rather than a state of being.

Any examination of the key point of 'menstrual modernity' around 1920 must note a confluence of factors. The body discussed by hygiene discourse was widely seen as properly functioning, disabled only in the sense of proneness to external attack, and the rise of the disposable everyday-use 'menstruation bandage' from 1893 in the UK was concur-rent with the transfer in Western popular medical thought to germs as the source of all disease following the work of Louis Pasteur and Joseph Lister in the 1860s and Robert Koch's founding of bacteriology in the 1870s.[10] The work of women doctors in both the US and the UK was also keenly instrumental in attacking ideas of menstrual debility by the 1920s.[11] The language of visual advertising for the first mass-produced disposable sanitary towels must also be recognised as a factor.[12] The menstrual state of the film vampire, however, indicates an early element of this transition to widespread notions of menstrual health and the contemporary modern menstrual/female body.

Notes

1. Kipling's short six-verse poem begins:
   > A fool there was and he made his prayer
   > (Even as you and I!)
   > To a rag and a bone and a hank of hair

(We called her the woman who did not care),
But the fool he called her his lady fair
(Even as you and I!)

Oh the years we waste and the tears we waste
And the work of our head and hand,
Belong to the woman who did not know
(And now we know that she never could know)
And did not understand. (220; emphasis in original)

2. McGee points out that the whereabouts of the painting have been unknown since 1903 (15).
3. For an example, see Higashi 58–9.
4. See Munford, this collection.
5. See Mulvey-Roberts and Munford, this collection.
6. As Cathy McClive points out in this book, 'menstruation' as a process – as opposed to 'menstrual blood' as a substance – only became the standard medical method of conceiving of the phenomenon in the West in the mid-eighteenth century. This lexical change reflected the growing understanding of holistic purpose that would give rise to physiology. Furthermore, given physiology's dedication to discovering the truth of every phenomenon in terms of its function (its departure from anatomy), its institutionalisation from its emergence in the middle of the nineteenth century to its dominance by the end, with the development of instruments such as the microscope in histology and the stromuhr for measuring blood flow, the establishment of chairs of physiology at University College London and Cambridge and Oxford Universities, the professional specialisation of physiologist, the establishment of physiology laboratories, the launch in 1878, of the *Journal of Physiology*, the proliferation of group-edited student textbooks utilising new print technologies, and the establishment of the Physiological Society in 1876 and the American Physiological Society in 1887, was almost certainly going to lead to the 'discovery' of the purpose of menstruation.
7. The term was not used in the first edition of Herbert Spencer's *Principles of Psychology* in 1855 but was added to the second edition in 1870 (142).
8. See Blackman, this collection.
9. This may well have been why the vampire became such a major figure of film discourse. Just as its encapsulation of vilified public female existence was recalled in a cultural and consumer context in which non-reproductive public femaleness was being commonly reinterpreted, its encapsulation of menstrual morbidity was recalled in the context of cinema's representational revisions of the nervous female body.
10. See Brumberg 113–14.
11. See Strange 622–3.
12. See Vostral, this collection.

Works cited

Allbutt, T. Clifford. 'Neurasthenia'. *Diseases of the Brain and Mental Diseases.* 1910. Vol. 8 of *A System of Medicine.* Ed. T. Clifford and Allbutt and Humphry Davy Rolleston. 2nd ed. 11 vols. London: Macmillan, 1905–11. 727–91.

Browne, Porter Emerson. *A Fool There Was. A Story.* 1907. London: Greening & Co., 1911.

Brumberg, Joan Jacobs. ' "Something Happens to Girls": Menarche and the Emergence of the Modern American Hygienic Imperative'. *Journal of the History of Sexuality* 4.1 (July 1993): 99–127.

Carpenter, W. Benjamin. *Principles of Human Physiology, with their Chief Applications to Psychology, Pathology, Therapeutics, Hygiene and Forensic Medicine.* 5th ed. London: J Churchill, 1855.

Chapman, John. *Functional Diseases of Women. Cases Illustrative of a New Method of Treating Them Through the Agency of the Nervous System by Cold and Heat.* London: Trübner & Co., 1863.

Codd, Elsie. 'Film Types 2: The Vampire'. *Pictures and the Picturegoer* 13.187 (8–15 September 1917): 301–2.

'A Few Simple Lessons in the Art of Becoming a Picture Player'. *Pictures and the Picturegoer* 6.20 (4 July 1914): 451.

Foster, Michael. 'Physiology'. 1885. Vol. 19 of *Encyclopedia Britannica*. 9th ed. 25 vols. Edinburgh: Adam and Charles Black, 1875–89. 8–23.

Foucault, Michel. *The Will to Knowledge.* 1976. Vol. 1 of *The History of Sexuality.* 3 vols. 1976–84. Trans Robert Hurley. London: Penguin, 1981.

Flint, Austin. *Treatise on the Principles and Practice of Medicine.* 3rd ed. Philadelphia: Henry C Lea, 1868.

A Fool There Was. Dir. Frank Powell. Perf. Theda Bara. Fox, 1915.

Geddes, Patrick. 'Reproduction'. 1882. Vol. 20 of *Encyclopaedia Britannica*. 9th ed. 25 vols. Edinburgh: Adam and Charles Black, 1875–89. 407–32.

'The Girl on the Film. No. 10: Miss Alma Taylor'. *Pictures and the Picturegoer* 6.20 (4 July 1914): 442–3.

'The Girl on the Film. No. 11: Miss Elisabeth Risdon'. *Pictures and the Picturegoer* 7.35 (17 October 1914): 150–2.

Gore, Ivan Patrick. 'A Little Earnest Thought on The She-Villain'. *Pictures and the Picturegoer* 18.321 (10 April 1920): 372.

'Here and There in Screenland'. *Pictures and the Picturegoer* 13.194 (27 October 1917): 473.

Higashi, Sumiko. *Virgins, Vamps and Flappers: The American Silent Movie Heroine.* Montreal: Eden Press, 1978.

'Kathlyn Williams: The Daring Star of the Selig Company'. *Moving Picture Stories* 2.27 (4 July 1913): 31.

Kipling, Rudyard. 'The Vampire'. 1897. *Rudyard Kipling's Verse. Definitve Edition.* 1940. London: Hodder & Stoughton, 1986. 220–1.

Laycock, Thomas. 'On the Reflex Function of the Brain'. *British and Foreign Medical Journal* 19 (1845): 298–311.

——. *A Treatise on the Nervous Diseases of Women; Comprising an Inquiry into the Nature, Causes and Treatment of Spinal and Hysterical Disorders.* London: Longman, Orme, Brown, Green and Longmans, 1840.

McGee, Tim. 'Sir Philip Burne-Jones: "A Life in a Tall Shadow." ' *Review of the Pre-Raphaelite Society* 7.3 (Autumn 1999): 7–20.

'Miss Dorothy Foster'. *The Pictures* 3.72 (February 1913): 13.

Mosucci, Ornella. *The Science of Woman: Gynaecology and Gender in England 1800–1929.* Cambridge: Cambridge University Press, 1990.

Mulvey-Roberts, Marie. 'Dracula and the Doctors: Bad Blood, Menstrual Taboo and the New Woman'. *Bram Stoker: History, Psychoanalysis, and the*

Gothic. Ed. William Hughes and Andrew Smith. New York: MacMillan, 1998. 78–95.

Naldi, Nita and La Marr, Barbara. 'This Business of Being a Vampire'. *Motion Picture* 29.2 (March 1925): 42–3.

'News and Notes'. *Pictures and Picturegoer* 14.209 (9–16 February 1918): 148.

'Notes and News'. *Pictures and Picturegoer* 18.318 (20 March 1920): 282–4.

Oliver, James. 'Menstruation – Its Nerve Origin – Not a Shedding of Mucous Membrane'. *Journal of Anatomy and Physiology* 21 (1887): 378–4.

Ord, William. *On Some Disorders of Nutrition Related with Affections of the Nervous System (Neurotic Dystrophies).* 1884. London: Harrison & Sons, 1885.

'Picture News and Notes'. *Pictures and the Picturegoer* 8.61 (17 April 1915): 38.

'Picture News and Notes'. *Pictures and the Picturegoer* 11.137 (30 September 1916): 2.

'Picture News and Notes'. *Pictures and the Picturegoer* 11.151 (6 January 1917): 314.

'Review of *A Fool There Was*'. *New York Dramatic Mirror* (20 January 1915): 4.

Sanatogen. 'Which Have you Got – Strong Nerves or Wrong Nerves?' Advertisement. *Pictures and Picturegoer* 17.296 (18 October 1919): 480.

Simpson, Alexander. *Lecture Introductory to a Course of Lectures on Midwifery and Diseases of Women and Children.* Edinburgh: Oliver & Boyd, 1870.

Spencer, Herbert. *The Principles of Psychology.* 2nd ed. 2 vols. London: Williams & Norgate, 1870–72.

Stoker, Bram. *Dracula.* 1897. London: Penguin, 1993.

Stephenson, William M. *On the Menstrual Wave.* New York: William Wood & Co., 1882.

Strange, Julie-Marie. 'Menstrual Fictions: Languages of Medicine and Menstruation, c. 1850–1930'. *Women's History Review* 9.3 (Spring 1996): 607–28.

Tait, Robert Lawson. 'Menstrual Irregularities and Their Relation to Diseases of the Nervous System'. *Obstetrical Journal of Great Britain and Ireland* 1 (1873): 94–104; 173–80.

Tilt, Edward John. *The Change of Life in Health and Disease: A Practical Treatise on the Nervous and Other Affections Incidental to Women at the Decline of Life.* 2nd ed. London: Churchill, 1857.

'Vampires: "Pictures" Page for the Fair Sex'. *Pictures and the Picturegoer* 11.159 (3 March 1917): 483.

Young, James Carruthers. 'Hysteria and Functional Nervous Diseases, with Special Reference to the Vaso-motor Element in the Aetiology'. MD Diss. University of Durham, February 1914.

18
Masking Menstruation: The Emergence of Menstrual Hygiene Products in the United States

Sharra L. Vostral

Managing menstrual periods by purchasing menstrual hygiene products is a thoroughly modern phenomenon. Throughout the nineteenth century, women in the United States hand-produced sanitary napkins, folding and sewing cotton, gauze, or rags into pads that could be pinned into their undergarments, laundered and reused each month. The practice of layering petticoats and the use of devices such as rubber aprons and bloomers helped to protect the outer garment from stains (Bullough 625). By the end of the First World War, medical practitioners viewed these menstrual rags as unhygienic. Not only did they catch blood, they argued, they trapped the 'decidedly offensive odor' of decomposing blood (Robinson 51–2). Hand-washed and hung out to dry, they could never be completely disinfected. Dr William Robinson, author of *Sex Knowledge for Women and Girls* (1917) warned women about the dangers of menstrual rags. He believed that they should not be used unless they were 'recently washed and kept wrapped up and protected from dust' (52). Unclean rags led to infection, and he had 'no doubt that many cases of leucorrhea [white discharge from the vagina] date back their origin to unwashed rags' (52). In order to reassert this point, he believed that 'hygiene of menstruation can be expressed in two words: cleanliness and rest' (52).

Although some companies attempted to sell manufactured disposable products in the late-nineteenth century, there was little cultural demand for them. However, following the First World War many things changed. The passage of the Nineteenth Amendment to the constitution of the United States in 1920 guaranteed women the right to vote and provided many women a sense of legitimacy as full citizens. This empowerment instilled confidence and demands for more than the vote. The introduction of the ill-fated Equal Rights Amendment (1923) was partly the result of this pressure. Cultural shifts occurred concurrently with these political changes. More women attended college, lived away from their families, and identified as single women. Part of this identity included consumption of manufactured products, including ready-made foods, clothing, cosmetics and cleaning products. Fuelled by the new knowledge that germs transmitted diseases, personal hygiene became a national preoccupation. The desire of the 'New Woman' to become modern through mass consumption provided the environment in which manufactured menstrual hygiene products could be purchased and used. By using store-bought menstrual hygiene products, white middle-class women were encouraged to leave home, go to school, work at the office, and travel in cars without discomfort getting in the way. Disposable sanitary pads meant that women could challenge the common wisdom that they remain homebound during their periods. Menstrual hygiene products enhanced the daily activities, work schedules, dress styles and new freedoms demanded by the New Woman of the 1920s who used them (Bullough 614).

Manufactured menstrual hygiene products marked a new aesthetics of waste and a new attention to the female body. An appeal to women as managers of menstruation marked a shift in thought from participating in mandatory rest, or even acting ill, to a practice of technologically masking menses. Since menstruation had been understood as a condition, ailment, disability, and altered emotional state, both men and women often treated menstruating women differently during their periods, ascribing poor performance, mental instability, and general irritability to a woman's period. However, using and purchasing sanitary napkins marked a clear break with conventions regarding women as temporarily invalid. If women could hide the blood and feign their non-menstruating bodies, they therefore might also avoid associated presumptions of instability. Disposable sanitary napkin advertisements beckoned women to fulfil their modern roles through emancipating their bodies: modern women's bodies required health, and menstrual hygiene products diminished the liability of menstruation because they provided a 'technological fix' to a social problem. Menstruation's 'hygienic handicap', to use the language of the first Kotex advertisements, could be managed if women purchased the proper sanitary napkins (Kotex, 'Woman's Greatest Hygienic Handicap'; see Figure 18.1). Sanitary napkins also helped women to 'pass' as normal, thus masking debility and presenting the appearance of a healthy female body.

Liberty   *December 31, 1927*

Woman's Greatest Hygienic Handicap

As Your Daughter's Doctor Views It

Easy Disposal
and 2 other important factors

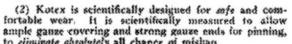

① Disposed of as easily as tissue. No laundry.

② True protection—5 times as absorbent as cotton.

③ Obtain without embarrassment, at drug or dry goods stores,* simply by saying "Kotex."

Because of the utter security this new way provides, it is widely urged by physicians—ABSOLUTE SECURITY, plus freedom forever from the embarrassing problem of disposal

By ELLEN J. BUCKLAND, Registered Nurse

SIXTY per cent of many of the commoner ailments of women, according to many medical authorities, are due to the use of insanitary, makeshift ways in meeting woman's most distressing hygienic problem.

For that reason, this new way is widely urged today. Especially in the important days of adolescence. On medical advice, thousands thus started first to employ it. Then found, besides, protection, security and peace-of-mind unknown before. Modern mothers thus advise their daughters—for health's sake and immaculacy.

Protection women KNOW is real

Kotex is more than a "sanitary pad"—it is scientific protection in the full sense of the term.

(1) Kotex is the only sanitary pad in the world today filled with *Cellucotton* wadding, the super-absorbent of modern scientific attainment. Thus Kotex absorbs *16 times its own weight* in moisture! Thus Kotex is 5 times as absorbent as the ordinary cotton pad!

Supplied also through vending cabinets in rest-rooms by West Disinfecting Co.

(2) Kotex is scientifically designed for *safe* and comfortable wear. It is scientifically measured to allow ample gauze covering and strong gauze ends for pinning, to *eliminate absolutely* all chance of mishap.

(3) Kotex actively *deodorizes*. Years of scientific research were spent in developing this obviously important factor.

No laundry

Kotex, too, ends for all time the embarrassing problem of disposal. One uses it, then discards it—*as easily as tissue.*

Ask for them by name—Kotex

"Genuine Kotex" is plainly stamped on every box. If offered a substitute said to be "like Kotex"—beware. Imitations are, we are told, being offered for the sake of higher profit by some stores, as the "same as Kotex." They are not. *Watch out.*

Only Kotex itself is "like" Kotex. Obtain at any store in boxes of 12 pads. Two sizes, Kotex Regular and Kotex-Super. Eight in 10 better-class women, throughout America, employ this unique and *certain* protection.

Kotex Company, 180 North Michigan Ave., Chicago, Ill.

"Ask for them by name"

KOTEX
PROTECTS—DEODORIZES

No laundry—discards as easily as a piece of tissue

Figure 18.1   By invoking the authority of a female physician and fictitious Nurse Ellen Buckland, Kimberly Clark Corporation relied upon a medicalised view of menstruation to promote Kotex sanitary napkins. The advertisement conveyed the sense that the technology of sanitary napkins would allow women to overcome the liability and ailment of menstruation. 'Woman's Greatest Hygienic Handicap'. *Liberty* 4.52 (31 December 1927).

Companies invested in the concept of debility in order to tout the transformative power of the product in creating sophisticated, modern, and functional menstruating women. The advertisements in effect repackaged and appropriated feminist sentiments while downplaying feminism's more disruptive elements. Products helped to define elements of modern womanhood without revolutionising gender relations.

That the sanitary napkin became a store-bought product is not surprising, but there were many developments that led to its acceptance as a consumer item. The rise of corporations and mass consumerism at the beginning of the twentieth century directly influenced the manner in which people from the United States conducted their daily activities. Although mass-produced products were not new, the 1890s saw a revolution in distribution infrastructures in the United States. New technologies of refrigeration, mechanisation for cardboard boxes and canning, and transportation systems provided the means for new products to be mass-distributed quickly. Local grocers easily purchased wholesale goods manufactured in states thousands of miles away. Companies encouraged brand name recognition, relying on the hope that consumers would choose neatly packaged Quaker Oats, for instance, over bulk stock sold in one hundred pound burlap bags (Strasser 170). The potential to carve out new consumer markets spurred production and encouraged consumption. Coupled with the rise of consumerism, mass advertising fuelled brand-name recognition (Lears 196). Companies discovered that consumers who trusted a product, relied upon a standard delivery and quality, and believed in its usefulness over another similar product were willing to pay a bit more for it. In turn, once consumers knew to shop for and purchase more expensive brand name products, advertisers warned them to 'accept no substitutes' (Graham 103). As food production and sales increased, items such as soap and other personal hygiene products also found lucrative markets. Through its ad campaign, the Prophylactic Toothbrush Company aimed to educate the public about benefits of maintaining oral health. Colgate published pamphlets about dental hygiene and shaving techniques. Gillette claimed that its razors provided a public service. A liberal gargling with Listerine mouthwash tackled the annoyance of bad breath and killed bad germs. Regular bathing and lathering with Ivory soap conquered body odour and washed it down the drain (Strasser 95).

The success of the products stemmed from attention to cleanliness, and an obsession about living in a germ-free environment (Hoy 133). Bathroom and kitchen design reflected the growing paranoia that germs spread disease and, as a result, easily disinfected white ceramic tile, enamel tabletops and porcelain sinks became immediate necessities. Hygiene was linked to the notion of civilisation: in 1917, *House and Garden* claimed that '[t]he bathroom is an index to civilization. Time was when it sufficed for a man to be civilized in his mind. We now require a civilization of the body. And in no line of house building has there been so great progress in recent years as in bathroom civilization' ('Bathrooms and Civilization' 90). Medical practitioners, social reformers, home economists and female consumers took the idea that cleanliness indicated advanced civilisation quite seriously. A clean, well-managed home both signified a high quality of life and a successful, prosperous family and promoted individual vigour and health. The body itself also required attention to hygiene in order to maintain vitality. In particular, control and regulation of bodily by-products helped to civilise the body and elevate it from its most banal processes. The fixation on cleanliness embodied elements of racial hierarchy and rationalisation of white, upper-class superiority; a clean, fresh smelling, perfectly coifed upper-middle class body looked and smelled very different from an unbathed, unkempt and working-class body. Within the model of perfectible civilisation, cleanliness represented human progress and indicated a level of personal control and bodily efficiency (Lears 165). This demand for clean and hygienic bodies raised an uncomfortable contradiction between public and private. Intimate matters of hygiene affected outcomes of public health and companies quickly capitalised on these fears (Lears 137). Advertisements for deodorant, talcum powder, toothpaste, laxatives, pomades, and breath fresheners broadcast nationally the most seemingly private moments of bathroom routine, thus subtly altering the relationship between person and body (Lears 165). Menstrual hygiene provides one of the best examples of the way women construed their physical experiences through public discourse about private bodily functions. The discourse was so significant that the process of menstruation became primarily understood as a matter of civilisation and a modern hygienic problem in the early-twentieth century (Brumberg, ' "Something Happens to Girls" ' 99; *The Body Project* 55).

Kimberly Clark Corporation led the way in disposable menstrual hygiene products with the introduction of Kotex sanitary napkins. In response to the reports of resourceful battlefield nurses that one of its products – cellucotton, a new bandage material manufactured from cellulose wood pulp and used as surgical dressing for wounded soldiers – worked well absorbing menstrual fluid, in 1919 Kimberly Clark formed a separate division – the Cellucotton Products Company – to sell a disposable sanitary napkin under the brand name Kotex, a contraction of

'COtton-like *TEX*ture' (Spurgeon 17). Advertisements first appeared in 1921 in *The Ladies' Home Journal*. By 1923, entire campaigns accompanied distribution of products to drugstores (Spurgeon 19). Since menstruation was still considered a 'condition' at best, it made sense that druggists sold sanitary pads along with other remedies. Women either phoned in an order or approached the clerk and asked for it by name. Similar to the purchase of Preparation-H to treat haemorrhoids, the 'code-name' meant that there was no need to publicly announce that ailment. Like Colgate toothpaste, early sanitary napkin advertisements used the rhetoric of hygiene to appeal to customers. One advertisement stated that 'Kotex are hygienic, convenient, and so low in cost that they form a new sanitary habit' ('Meets the most exacting needs' 123). Like the Gillette razor and Prophylactic toothbrush, Kotex was disposable and made to be discarded. Advertisements announced that 'Kotex is cheap enough to throw away and easy to dispose of by following simple directions in every box' ('Traveling or at home' 118). Once used, Kotex could be burned or thrown away, without guilt or the hassle of laundering. According to Miss Harris, the head nurse in the dispensary of the Metropolitan Life Insurance Company, Kotex could be flushed down the toilet 'if the gauze is taken off, the filler torn apart and allowed to soak before being flushed' (Gilbreth, 'Report' 20–1). Though an awkward task, disposing the products promised to fulfil the quest for a germ-free environment and a clean body. The early advertisements employed medical authority to create the presentation of a legitimate, salubrious and hygienic product (Marchand 355). A fictitious character, graduate Nurse Ellen Buckland, instilled fears that makeshift pads were not only unsanitary but unsafe as well. Her one-to-one conversational tone encouraged a more personal relationship with the product while her vague references to medical claims created the tone of expertise. Advertisers believed that the copy educated consumers and that the empathetic yet stern female 'graduate' Nurse Buckland urged women to heed her advice.

Use of menstrual hygiene products supposedly created a sanitary habit that translated into a civilised body (Kane 137). The products helped to transform bodily excretions into an appropriate behaviour, all by linking hygiene to upward, white mobility. Kotex advertisements projected refined images of socially mobile and financially secure women. One advertisement read, '[l]ow price sometimes causes people of means and refinement to hesitate in buying a new article. Kotex is inexpensive, yet women who can afford the best were first to accept Kotex' ('Traveling or at home' 118). These advertisements targeted moneyed, well-educated white women, in the belief that they would be willing to pay the high price of five cents per pad (equivalent to a loaf of bread). The success of the sale relied upon women's willingness to learn about and experiment with a new product, both of which were invoked as class-elevating attitudes. Mindful of changing roles and requirements for bodily freedom, the first Kotex advertisements, designed by Wallace Meyer, appropriated and reinterpreted aspects of feminism. The products claimed to offer to women a realm of wholly unknown freedoms – of movement and travel, and from leakage and laundry – available only through using Kotex: '[t]hey are unsurpassed in business, when travelling, or at home' ('It forms a new sanitary habit' 184). As historian Nancy Cott argues, advertisers reworked feminist agendas and translated them so that women's purchases exerted the appearance of choice and control (174). In effect, this collapsed women's demand for economic, marital, and educational choice into consumerism. Possessing the power to choose a product supposedly led to personal fulfilment. By exercising their right to choose Kotex, they appeared to liberate and modernise their bodies from the ideological and physical restraints of the pre-modern menstrual rag. Menstrual hygiene advertisements cleverly appropriated this technique to turn the purchase of sanitary napkins into an act of personal fulfilment and emancipation. One advertisement claimed that 'the comforts of Kotex' played a vital part 'in new-found feminine freedom and peace-of-mind' ('Other women will tell you' 53).

Part of this peace of mind for women came from the ability to hide menstruation and to, in effect, 'pass' as normal. The use of menstrual hygiene products allowed a certain class stratum a degree of control in a world still willing to judge menstruation as a liability. Kimberly Clark Company participated in this masking, with advertisements assuring women that sanitary napkins could be purchased hassle-free: 'Kotex saves embarrassment in many ways: It is easy to buy without counter conversation by asking not for "sanitary pads" but "Kotex" ' ('Wherever nice women gather' 128). Of course, the name became nationally recognised, but the system of masking persisted. Kotex advertisements claimed to hedge embarrassing situations because of simple packaging: 'a plain blue box free from descriptive matter, the name the only printing' ('Wherever nice women gather' 128). Though consumers knew exactly what was in the box, advertisements implied that a woman could purchase needed supplies while maintaining discretion. Eventually, the simple blue box easily symbolised Kotex – and thus menstruation – and required further masking to assure propriety. Many local grocers took the extra step of wrapping Kotex boxes in newspaper to quell fears of embarrassment. In the autumn of 1926, Mrs V.V. Davidson of Charleston, West Virginia wrote to Johnson & Johnson who, at the time, manufactured Nupak sanitary napkins. Wanting to offer advice about its products and those of its competitors, she hoped to be compensated for her ideas. She complained that the size and shape of both the Nupak and Kotex sanitary napkin boxes were too obvious. The problem with Kotex was that it was 'so well advertised that to see a woman carrying the package is to know what it contains and, logically, the condition she is in' (Davidson). The blue box and rectangular shape, she argued, could not even be masked when wrapped in newsprint. 'Despite the fact we are getting over a lot of our modesty,' Davidson continued, 'that is one condition we do not like to have exposed to the public in general as we do when we walk down the street carrying your well known package' (Davidson). Johnson & Johnson would address this problem in part with the 1928 'silent purchase coupon' for Modess sanitary napkins, first created in 1926. Printed so that 'Modess may be obtained in a crowded store without embarrassment or discussion,' it ostensibly allowed a woman to simply hand over a coupon which requested 'one box of Modess, please' ('Silent Purchase' 135; see Figure 18.2).

The silent coupon may not have helped much since problems extended beyond the purchase with the cashier. At Smith College, one observer noted that while students 'chastely' hid Kotex, 'the size and shape are so well known that anyone carrying a box frequently comes in [to the dormitory] for much humor and taunting and this is accordingly not a pleasant feature of the napkin' (Gilbreth, 'Report' 70). Along with being bulky and oversised, it became increasingly clear that the newly marketed menstrual hygiene products were not very good. They bunched up, leaked, and chafed women's thighs. Many preferred the old laundered rags to these poorly performing, semi-disposable napkins. This was bad news for Johnson & Johnson who already had an unsuccessful track record with menstrual hygiene products. In 1896 Johnson and Johnson had manufactured the first mass-produced sanitary napkin, Lister's Towels – disposable, gauze-covered cotton pads. Sold directly through the Sears and Roebuck catalogue, the towels became lost on pages filled with dubious claims and small sketches of products. Without a major promotional campaign, and luke-warm word-of-mouth advertising, Lister's Towels hobbled along in market sales (Delaney *et al.* 138). Because Kotex displayed enormous market expansion, competitors quickly followed. Cautious to re-enter the market due to the past poor performance of Lister's Towels and abysmal sales of

251

June, 1928   LADIES' HOME JOURNAL   135

Silent Purchase
A Modess
Advantage

AT LAST, a silent purchase plan for a sanitary napkin! And a napkin of wonderful improvements—
soft, protective, absolutely disposable—Modess.
Women are buying it, marveling at it, and buying again. But even with this new and ideal napkin
the old embarrassment of purchasing was still a
problem until Johnson & Johnson solved it so easily
and ingeniously that you will wonder no one thought
of it before.

In order that Modess may be obtained in a crowded
store without embarrassment or discussion, Johnson
& Johnson devised the Silent Purchase Coupon presented below. Simply cut it out and hand to the
sales person. You will receive one box of Modess.
Could anything be easier? Is there a woman anywhere who will not be grateful for this method of
silent purchase?

Your first Modess will be a revelation of unhoped-
for comfort. The great Johnson & Johnson laboratories worked four years to make the finest and most
comfortable sanitary napkin ever offered to women.
An entirely new substance, soft as the finest cotton,
was invented for the disposable center. The gauze is
specially softened and sides are gently rounded to
prevent chafing. Modess has a moisture-resisting
back that makes the comfort it bestows doubly
imposing a sanitary product of amazing superiority.

Fifty cents for a box of twelve.

Modess
so infinitely finer

**SILENT
PURCHASE
COUPON**

To Sales Person—
One box of Modess, please
The New sanitary napkin made by
Johnson & Johnson
NEW BRUNSWICK, N.J. U.S.A.

**SIX
SUPERIORITIES**

1. Gauze specially softened
with a film of down.

2. Pliant fluffy filler of
amazing absorbency.

3. Rounded sides assuring
comfort and no
clumsiness.

4. A moisture-resisting back
giving positive security.

5. Disposable—flushes away.

6. Silent purchase coupon.

© by J & J

Figure 18.2   135 Johnson & Johnson devised the 'silent purchase coupon' so that
no potentially embarrassing discussion of menstruation or sanitary napkins
ensued between the menstruous woman and the salesperson. 'Silent Purchase',
Ladies' Home Journal 45.6 (June 1928).

Nupak, Johnson & Johnson chose to approach the menstrual hygiene arena armed with data. According to its market research, Kotex cornered at least 75 per cent of all sanitary napkin sales; Nupak held less than 2 per cent of the market, and Lister's Towels 0.006 per cent (Gilbreth, 'Report' 62).[1] Although Johnson & Johnson had usually relied upon customer's letters and the experiences of their female employees for feedback, they needed much more precise information about increasing market share (Graham 217). Broad-scale market research was not the standard, but Johnson & Johnson was anxious to tap into the wealth of female consumers. R.W. Johnson hired efficiency expert Lillian Gilbreth as a consultant for its new but faltering product Modess sanitary napkins, launched in 1926.

Gilbreth had established herself as a pre-eminent expert in time-motion studies, gaining fame with her husband Frank as co-director of the management consulting firm Gilbreth, Incorporated. They primarily analyzed repetitive motions within the workplace and ascertained the most efficient procedures for management to implement in their indus-trial factories. Her educational background and PhD in psychology aided the pair by incorporating human behaviour with mechanical engineering to study 'motion psychology'. Although Frank died in 1924, Lillian continued with the company and was productive with their work well into the 1960s. She served on special committees during the Hoover, Roosevelt, Eisenhower, Kennedy and Johnson administrations, providing advice on issues from civil defence, war production, ageing, and rehabilitation of physically disabled people. She was the first woman to gain membership in the American Society of Mechanical Engineers in 1926, and was also the first female professor in the engi-neering school at Purdue University in 1935. She considered herself an expert in human psychology, industrial management, efficiency models, business consultation and home economics, later referring to her per-sonal expertise earned by rearing twelve children (Gilbreth, *As I Remember* 173). As a proponent of scientific management, Gilbreth believed there was indeed one superior method to accomplish any given task. The techniques which she applied to time-motion studies in factories, she believed, were easily applicable to the home as well. Gilbreth argued that if educated women applied scientific management to the home and made use of manufactured items, they would gain more time for leisure and other life pursuits. Through home economics and the systematic application of efficient technology, women could more professionally manage their homes. This efficiency involved purchasing manufactured goods.

Newly widowed and with few clients interested in her as a lone female researcher, she redefined her services to study women's consumer behaviour (Graham 85–105). When R.W. Johnson presented her with the opportunity to design, produce, and market 'a new type of sanitary napkin to supplement those already made by Johnson and Johnson' she approached the project with keen interest and vigour (Gilbreth, 'Report' 9). For a one-time fee of $6,000, she conducted the most thorough marketing research on sanitary napkins and pads up to that time, submitted in 1927 simply as 'Report of Gilbreth, Inc'. Gilbreth's interest in efficiency models easily translated onto women's bodies: if manufacturing processes could be streamlined, so too could elimination of waste from the body. In this regard, the first goal she had for the project was to apply social engineering and 'determine what types of napkins would be most serviceable' (Gilbreth, 'Report' 1). Assuming menstruation to be a detriment, her second objective was to study 'the period during which sanitary napkins are used with the hope of making a contribution to the health problem' (Gilbreth, 'Report' 1). Developing an efficient sanitary napkin for women, she believed, would provide them control and agency through better body mechanics during menstruation, and offer women better tools to manage debility. By purchasing sanitary pads, women could maximise efficiency of the body and minimise menstrual angst. Gilbreth worked within and accepted the premise of debility, and used scientific methods of observation and data collection to help manage, but not necessarily to revolutionise, the body. She convincingly justified the need for a systematic study of menstruation. She estimated that in 1926 there were 'approximately 30,000,000 women between the ages of 13 and 45' in the United States who menstruated. She assumed that the average woman menstruated 13 times per year, and that her menstrual life extended 32 years, amounting to 416 periods during a lifetime. She calculated that a woman used eleven pads each period, equating to 4576 napkins during a woman's menstrual life. She figured that if even one-third of the market were saturated, which it was not, that came to over 45 billion napkins to be produced. This indicated a market entirely ripe for exploitation (Gilbreth, 'Report' 15).

Gilbreth and her two staff members evaluated 53 different sanitary napkins on the market, including those marketed in the United States, Canada, and England with their accompanying belts, pins, clasps, rubber aprons and 'step-ins' – waterproof underwear worn between the sanitary pad and a woman's clothing. From the examination they hoped to determine 'what type of equipment is ideal to meet the needs of the period'. Gilbreth began with the assumption that 'all existing equipment is probably wrong' because, 'copied from a home made product or from a hospital pad designed for obstetrical use', they were 'not designed to meet the actual needs' of menstruating women (Gilbreth, 'Report' 9–10). Although the detailed examination of disposable products and menstrual equipment provided useful data concerning product shortcomings, the report to Johnson & Johnson also incorporated a revealing field survey. Gilbreth and her staff contacted more than twenty groups to partake in the survey, including college instructors at the 'Big Four' (Smith, Radcliffe, Vassar, and Wellesley), the Women's Bureau Department of Labor, the National Committee for Prison Work, Winchester Laundries, Public Schools, teacher training Normal Schools, and Medical Schools. She limited the study to young women and the 'college girl' even though they represented the minority of menstruating women. She admitted that '[w]hile it may seem that the home making women are still in the majority, her needs are not typical of the most difficult needs to meet'. The college student proved to be more challenging because she 'sets the styles in clothing' and demands the most from sanitary napkins due to her activities (Gilbreth, 'Report' 12).

The survey that Gilbreth conducted sought 'to secure a comprehensive opinion on the requirements of an ideal sanitary napkin' (Gilbreth, 'Report' 14). She and her staff obtained data using three methods. They sent out 3,000 questionnaires, of which 1,037 were returned, held conferences at different colleges with specific target groups in which they discussed current products on the market and their shortcomings, and interviewed college presidents, deans, and faculty members of hygiene departments to ascertain their views concerning the products. Although many of the test groups she targeted chose not to participate, those that did displayed 'great interest ... in the possibility of a new sanitary napkin, and questions were frequent as to when it would appear, what name it would bear, and who was manufacturing it' (Gilbreth, 'Report' 14). If nothing else, it proved to the Johnson & Johnson Company 'the friendly interest and the potential market it has developed for their napkin' (Gilbreth, 'Report' 14). The survey asked questions such as 'What type of napkin do you use? How many each period? Per year?' They became increasingly detailed. 'Where do you buy your supplies? Why?' and 'Do you alter the napkin before you use it?' (Gilbreth, 'Report' 5). Of the 1,000 college and businesswomen surveyed, only 91, or 9 per cent, still used homemade napkins. Those who continued to make their own believed they could make them more comfortably, and of a better size than those found on the market. Gilbreth reasoned that girls living at home, with mothers who had materials on hand, usually constructed their own sanitary napkins. 'But the college and business woman, the majority of whom live in single rooms or small apartments have little space for materials that have to be kept about' and had little time to make them (Gilbreth, 'Report' 16). Of all the 1,037 women, only 16 per cent indicated that they were satisfied with either homemade or manufactured products. Gilbreth called this low figure 'significant' because it 'would seem to indicate to us something lacking in the commercial napkins now available' (Gilbreth, 'Report' 16). Yet collegiate and working women still purchased sanitary napkins, despite the drawbacks, because they afforded them a certain degree of physical freedom.

Within her questionnaire, Gilbreth assumed that women felt entirely too embarrassed to articulate the need for sanitary napkins to male clerks. She believed that women preferred female sales clerks and would seek out their assistance regardless of price of product or location. To her surprise, this was not the case. She found that, by a margin of two to one, women favoured low price, location of store, and the convenience of phoning in an order and charging it to an account over the service of a woman. Armed with this evidence, Gilbreth concluded '[i]t seems reasonable to suppose then from these general returns that if the desirable napkin is placed upon the market, women will have no hesitation in buying it' (Gilbreth, 'Report' 17). Since many products were of such poor quality, even the slightest improvement would boost sales. Gilbreth suggested that youthful women were the consumers who would most likely experiment with and purchase new products. She advised the advertising department 'to advertise to the girls themselves' in lieu of the older generation since 'in most families the daughters are the ones who undertake to do the telling' to their mothers (Gilbreth, 'Report' 70). Sanitary napkins were poignant artefacts representing a generational gulf. Upon Gilbreth's recommendation, a new series of advertisements emerged from Johnson & Johnson in 1929 called 'Modernizing Mother', installed in episodes as if a radio programme. In episode number three, Mother learns from Daughter how to do stretching exercises. She also discovers Modess. 'Millions of mothers whose girlhood was repressed are being trained by daughters to be young again – to know freedom – to grasp the idea that drudgery and useless labor are a sinful waste of life' ('Don't weaken Mother' 105). In episode number five, Mother learns to play golf. This time, the 'game' is not only golf, but also the 'game of escaping the bondage of old-fashioned ideas and being happily young again' ('Never mind, Mother' 87). These advertisements demonstrate how feminism, stripped of its more radical overtones, became coupled with cultural understandings of Modess and disposable sanitary napkins in general. The advertisements lauded the efficiency of the home and body, disparaged former methods of housework in favour of household mechanisation (which ironically increased attention to housework), and valued youthfulness as a worthy characteristic (Cowan 13–14). Older women using Modess could become modern and younger women could gain freedom – the generational difference between mother and daughter represented by managing the body with manufactured sanitary napkins.

The importance of Lillian Gilbreth to the sanitary napkin industry was that she provided the most detailed accounts to date of women's preferences in menstrual hygiene. By using carefully crafted surveys and statistical analysis, she provided far more than an estimated guess about women's likes and dislikes (which the company had relied upon previously), a market survey that quantified women's reactions. Johnson & Johnson drew upon the expertise of Gilbreth to design new products according to this social engineering. The study relied upon the underlying belief in debility, with the new and improved Modess pads providing a means to fix it. One woman liked the newly-improved smaller size of Modess so much that she attested that for the first time in her life she had 'not been uncomfortable through fear of conspicuousness' (Gilbreth, 'Report' 67). Modess repaired a body understood as liable to betray itself. However, many of the students informed Gilbreth of their desire for an entirely different disposable product, which truly concealed menstrual blood. 'What most of them wish, (and as yet they have not had this wish granted),' she noted 'is for a new product which will be completely invisible no matter how tight or thin their clothes are' (Gilbreth, 'Report' 70). This would offer the ultimate in personal and physical freedom from menstrual debility and the equipment of sanitary napkins. Sanitary napkins failed in this regard. Gilbreth recommended 'if a comfortable and satisfactory napkin were made available to consumers, even if it was the same price as Kotex, they would buy it' (Gilbreth, 'Report' 19). This provided space in the emerging menstrual hygiene industry for the new technology of tampons, patented in 1931 and developed by Tampax Sales Corporation in 1934.

The new menstrual hygiene products of sanitary pads and tampons more effectively concealed menstruation and helped to hide women's menstruating bodies. Literally bandaging the wound with scientifically developed dressings, the sanitary napkins concealed, controlled, and contained the blood better than menstrual rags. Some of the first market research, such as that conducted by Gilbreth, tapped into the lucrative market by addressing women's desires. However, their use and sale was still premised upon removing debility and embarrassment, and concealing the menstruating body. The success of Kotex, the challenges by Modess and the rising interest in an 'internal sanitary napkin' demonstrated women's desires for products that they liked and that helped them handle a process they understood as debility.[2] Although sanitary napkins are not usually linked with modernity, their design, manufacture, sale, and consumption marks a clear intersection of gender, feminism, and the modern woman. Menstrual hygiene products provided a technological fix to a bodily and social problem. Managing menses and therefore the temporary physical and mental debility seen to be associated with it, masking menstruating bodies and allowing women to 'pass' as normal, and providing women control and agency through better efficiency of the body, technologies of menstrual hygiene in the 1920s and 1930s changed the way that women understood their menstrual periods.

Notes

1. The survey results of 1037 women's preferences were as follows: Kotex (Kimberly Clark Corporation) 788; Curads (Lewis Manufacturing Company) 86; Venus (Venus Corporation) 36; Nupak (Johnson & Johnson) 20; Lister's Towels (Johnson & Johnson) 6.
2. Fax tampons, manufactured by the Sanitary Products Company in the mid 1930s, referred to its product on packaging and advertisement copy as an 'internal sanitary napkin'.

Works cited

'Bathrooms and Civilization'. *House and Garden* 31.2 (February 1917): 90.

Brumberg, Joan Jacobs. ' "Something Happens to Girls": Menarche and the Emergence of the Modern American Hygienic Imperative'. *Journal of the History of Sexuality* 4.1 (July 1993): 99–127.

——. *The Body Project: An Intimate History of American Girls*. New York: Random House, 1997.

Bullough, Vern. 'Merchandising the Sanitary Napkin: Lillian Gilbreth's 1927 Survey'. *Signs* 10 (Spring 1985): 614–27.

Cott, Nancy. *The Grounding of Modern Feminism*. New Haven: Yale University Press, 1987.

Cowan, Ruth Schwartz. *More Work for Mother: The Ironies of Household Technology from the Open Hearth to the Microwave*. New York: Basic, 1983.

Davidson, V.V. to Johnson & Johnson, 12 October 1926. Frank and Lillian Gilbreth Papers. Special Collections, Purdue University, West Lafayette.

Delaney, Janice, Mary Jane Lupton and Emily Toth. *The Curse: A Cultural History of Menstruation*. Urbana: Illinois University Press, 1988.

Gilbreth, Lillian M. *As I Remember: An Autobiography*. Norcross: Engineering & Management Press, 1998.

Gilbreth, Lillian. 'Report of Gilbreth, Inc.' 1 January 1927. Frank and Lillian Gilbreth Papers. Special Collections, Purdue University, West Lafayette.

Graham, Laurel. *Managing On Her Own: Dr. Lillian Gilbreth and Women's Work in the Interwar Era*. Norcross: Engineering & Management Press, 1998.

Hoy, Suellen. *Chasing Dirt: The American Pursuit of Cleanliness*. Oxford: Oxford University Press, 1995.

Kane, Kathleen. 'Feminine Hygiene Commercials: A Political and Symbolic Economy'. PhD. Diss. Northwestern University, 1992.

Kotex. 'It Forms a New Sanitary Habit Among Women'. Advertisement. *Ladies' Home Journal* 39.11 (November 1922): 184.

——. 'Meets the Most Exacting Needs'. Advertisement. *Ladies' Home Journal* 39.1 (January 1922): 123.

——. 'Traveling or at Home–Kotex is Almost Indispensable'. Advertisement. *Ladies' Home Journal* 39.7 (July 1922): 118.

——. 'Wherever Nice Women Gather'. Advertisement. *Ladies' Home Journal* 39.9 (September 1922): 128.

——. 'Other Women Will Tell You'. Advertisement. *Ladies' Home Journal* 46.4 (April 1929): 53.

——. 'Woman's Greatest Hygienic Handicap'. Advertisement. *Liberty* 4.52 (31 December 1927).

Lears, Jackson. *Fables of Abundance: A Cultural History of Advertising in America*. New York: Basic, 1994.

Marchand, Roland. *Advertising the American Dream: Making Way for Modernity, 1920–1940*. Berkeley: California University Press, 1985.

Modess. 'Don't Weaken Mother'. Advertisement. *Ladies' Home Journal* 46.4 (April 1929): 105.

——. 'Never Mind, Mother – You'll Learn'. Advertisement. *Pictorial Review* 30.9 (June 1929): 87.

——. 'Silent Purchase, A Modess Advantage'. Advertisement. *Ladies' Home Journal* 45.6 (June 1928): 135.

Robinson, William. *Sex Knowledge for Women and Girls: What Every Woman and Girl Should Know*. New York: The Critic and Guide Company, 1917.

Spurgeon, Anne M. 'Marketing the Unmentionable: Wallace Meyer and the Introduction of Kotex'. *The Maryland Historian* 18 (1988): 17–30.

Strasser, Susan. *Satisfaction Guaranteed: The Making of the American Mass Market*. New York: Pantheon, 1989.

19
Blood, Laughter and the Medusa: The Gothic Heroine as Menstrual Monster

Rebecca Munford

Bloody narratives

There has always been blood in the Gothic. A genre defined by exchanges, overspill and abjection, the Gothic's bloody excesses are, most often, negotiated over the site of the adolescent female body. While a sense of 'self-fear and self-disgust directed towards the female role, female sexuality ... and procreation' (Fleenor 15) has been identified as a central preoccupation of the 'female Gothic', the 'male Gothic' is overflowing with the bleeding bodies of its violated and murdered female protagonists.[1] In the Marquis de Sade's *Justine* (1791), for example, the flow of menstrual blood is repeatedly occluded as the 'virginal' state of the adolescent heroine is perpetually renewed and re-written in the blood of violent sexual initiation.[2] In the nineteenth-century decadent Gothics of Charles Baudelaire and Bram Stoker, the haemorrhaging bodies of the Sadeian imaginary are recast via the motif of the female vampire, the pre-eminent figure of the Gothic's bloody narratives. Indeed, female vampires have a particular presence in nineteenth-century poetry and fiction where 'they chiefly act as a warning against being taken in by appearances and becoming victim to the evils of women's active sexuality, equated with the demonic' (Wisker 169).[3] Incorporating the sex-drive and the death-drive, the female vampire (dis)embodies the simultaneous desire for, and dread of, female sexuality that is at the heart of the male Gothic.

By calling into question the delineation between inside/outside, I/Other, human/monster and living/dead, the vampire invokes the horrors of the in-between. Not only representing the uncanniness of the undead, the vampire is a figure of abjection.[4] Foregrounding the exchange of bodily fluids and the transgression of corporeal boundaries, the vampire embodies those 'impure' elements – urine, blood, semen, excrement, and so on – which, Julia Kristeva suggests, place the subject 'in perpetual danger' (9). In her schema of 'polluting objects', Kristeva distinguishes between the excremental and the menstrual. While '[e]xcrement and its equivalents (decay, infection, disease, corpse, etc.) stand for the danger to identity that comes from without: the ego threatened by the non-ego, society threatened by its outside, life by death,' menstrual blood 'stands for the danger issuing from within the identity (social or sexual); it threatens the relationship between the sexes within a social aggregate and, through internalisation, the identity of each sex in the face of sexual difference' (71). What is crucial about menstrual blood, then, is not merely its differentiation between male and female but, specifically, its foregrounding of 'the differences between men and *mothers* (or potential mothers)' (Grosz 92; emphasis in original). Embodying a dread of female sexuality and the 'generative power' of the archaic mother (Kristeva 77), the female vampire typifies the horror of sexual difference and the abject status of menstrual blood.[5] It is this notion of gendered blood – representing difference, disease, 'generative power' and death – which is (re)figured in the vampire narratives explored here. Mapping the iconographies of vampirism and vampirisation which have haunted the construction and circumscription of the female Gothic body since the nineteenth century, and which are exemplified by Bram Stoker's *Dracula* (1897), this chapter will consider the re-appropriation and re-working of the Gothic heroine as menstrual subject by contemporary French writer Pierrette Fleutiaux in *Histoire de la chauve-souris* (1975).

Vampires and ventriloquists

The burgeoning of female vampire iconography in the latter half of the nineteenth century was intricately related to medical discourses of the period and, in particular, the formalised hystericisation of the 'unruly' female body as, to use Michel Foucault's oft-cited account, 'analyzed – qualified and disqualified – as being *thoroughly saturated* with sexuality; whereby it was integrated into the sphere of medical practices, by reason of a pathology intrinsic to it' (104; emphasis added). Indeed, a supernatural nomenclature has haunted designations of hysteria for several millennia – from the theory of the 'wandering womb', chronicled as early as c. 1900 BCE in the Egyptian *Kahun Papyrus*, through the witch trials of the Middle Ages to Jean-Martin Charcot's acts of 'exorcism' at La Salpêtrière hospital in Paris during the 1880s. By this time, the vampire metaphor was prevalent in medical discourses of hysteria – and, in particular, menstruation – which represented woman as 'an incontinent slave to her secretions, unable to control her dripping, flowing, spurting, oozing bodily fluids' (Beizer 41). The nineteenth-century physician Oliver Wendell Holmes, for example, proposed that 'a hysterical girl is a vampire who sucks the blood of the healthy people about her' (qtd. Smith-Rosenberg 207), while contemporaneous constructions of nymphomania linked female lust to the need to replenish lost menstrual blood (see Showalter 180).

The menstrual body, then, is positioned as a hysterical body: a *dis-ordered* body that needs to be regulated, controlled and *contained*.

> Paradoxically, it is from the very moment a woman begins to menstruate, becomes *réglée*, regulated, fixed in her (female) role, that her entire being is open to *derangement*. ... It then seems that the specific preoccupation with menstrual disorders hides a more fundamental anxiety about menstruation *as* order, or the essential disorder of the female condition. (Beizer 40; emphasis in original)[6]

An example of late nineteenth-century medical attempts to 'fix' the unruly female body is Charcot's classification of, and experiments on, his female hysterical patients. Transforming the hospital space into his 'museum of living pathology', Charcot, during his (in)famous *leçons du mardi* (Tuesday lectures), fixed his female hysterical patients within a frame of erotic voyeurism, puppeteering grotesque public stagings of their psychic ailments for the scrutiny and analysis of a 'medical gaze'. In her discussion of 'Charcot's vampires', Elisabeth Bronfen describes how '[i]n Charcot's theatre of living pathology the two sides of the hystericized body could be read and reproduced endlessly: the corporal surface, skin, and gestures, poses and attitudes that drew the interpellating Other into the field of vision' (*The Knotted Subject* 176). The hysterical body is thus inscribed, deciphered and rendered 'readable', constructed through the doctor and patient's performance of a 'relationship of mutually dependent desires, gazes, and knowledge' (190). Moreover, as Bronfen highlights, Charcot's diagnostic methodology *'feeds off* a well-established, visual iconography developed in conjunction with notions of demonic possession and exorcism' (174; emphasis added). Janet Beizer's study of the staging of the hysteric's semiotic body within nineteenth-century literary and medical narratives highlights how the 'dermographic' patient in particular becomes an inscribed body that is not only spoken about, but *spoken for*. 'In fact the body does not speak; it is spoken, ventriloquized by the master text that makes it signify' (26). As the male physician writes on the body of the female patient, she is transformed into a *tabula rasa* bearing the imprint of his analysis, a blank text through which he 'ventriloquises' his own bloody narrative. Here, then, the female body is both vampiric in the threat it poses to the (male) subject, but also 'vampirised' through this act of ventriloquism.

This iconography of medical (dis)possession prefigures and provides an intertext for the literary representation of female vampirism in *Dracula*, where, as Marie Mulvey-Roberts has suggested, 'the interaction between the vampire and the victim … is a trope for the relationship between the Victorian male doctor and the female hysteric' (80). For, in its emphasis on the exchange, control and containment of women's bodies, *Dracula* is not so much an iteration of female *vampirism* as it is of female *vampirisation*. Moreover, while Stoker's text makes extensive reference to contemporaneous ideas about hysteria and its treatments – including the theories of hypnosis developed by 'the great Charcot' (Stoker 247) – the female vampire is most powerfully codified in relation to menstrual discourse. As Mulvey-Roberts proposes in her seminal essay on the novel,

> *Dracula* is, of course, far more than a novel about pathologies. … Its gendering of male blood as good and female blood as bad signals that it is menstrual blood and its pathologies that provoke a sense of horror. … Stoker's attention to the relationship between women and blood is a surrogate for menstrual taboo, which is also eroticised haemofetishism. At the same time, it is a reinforcement of the Victorian conservative medical view that menstruation should be morbidified. (78)

Indeed, Stoker's Gothic heroines, Lucy Westenra and Mina Harker, are primarily located as menstrual subjects in relation to the bloody bite of Count Dracula; by 'admitting' the Count, and breaking menstrual taboo, they are pathologised as menstrual monsters, forever on the verge of a personally, and *socially*, disabling hysteria. It is the 'dis-order' of menstruation that is regulated and censored by the 'Crew of Light', led by Van Helsing and Dr Seward as representatives of late nineteenth-century institutional medicine and science. Centrally, Lucy's positioning as menstrual monster is constructed through her identification with the Medusa, an alignment that is foregrounded by Dr Seward's description of her vampiric state: 'The beautiful colour became livid, the eyes seemed to throw out sparks of hell-fire, the brows were wrinkled as though the folds of the flesh were coils of Medusa's snakes. ... If ever a face meant death – if looks could kill – we saw it at that moment' (272). As Mulvey-Roberts highlights in this collection, the figure of the Medusa has frequently been associated with the menstrual female subject, the sight of whom was considered to be as deadly (or, rather, as *castrating*) as the gaze of the Gorgon.[7] Ultimately, then, Lucy's dangerous, disordered body must be destroyed. Notably, she is penetrated with a stake and then decapitated in an act of explicitly gendered violence that recalls Perseus' slaying of the Medusa (Stoker 277). However, at the same time as Lucy's vampiric state represents the threat posed by the menstrual subject, it is her *vampirisation* that marks her re-assimilation into and encasement within traditional marital frameworks. As Nina Auerbach highlights, as a vampire Lucy's desires are entirely focused on her fiancé Arthur so that she 'becomes more virtuous in death that she was in life. ... Lucy the flirt is purified into Lucy the wife' (79–80; see Stoker 272). It is only after Lucy's 'purity' has been reinstated that the 'boundaries between life and death, body and soul, earth and heaven' can be restored (Botting 151).

Although the characterisation of Lucy as Gothic heroine is most closely aligned with late nineteenth-century models of hysteria, the positioning of Mina as hysterical/menstrual subject also foregrounds a pertinent image of the ventriloquism explored earlier in this section. Centrally, Mina's menstrual identity is highlighted through one of the most disturbingly violent moments of Stoker's novel: her vampirisation by the Count. In this scene, the Crew of Light find Mina kneeling before Dracula, sucking the blood from the 'thin open wound' on his chest:

> Kneeling on the near edge of the bed facing outwards was the white-clad figure of [Mina]. By her side stood a tall, thin man, clad in black ... his right hand gripped her by the back of the neck, forcing her face down on his bosom. Her white nightdress was smeared with blood. ... Her face was ghastly, with a pallor which was accentuated by the blood which smeared her lips and cheeks and chin; from her throat trickled a stream of blood. (363–4)

The abject status of Mina's menstruating, contaminated and 'dirty' body is further signalled by her horrified cry – 'Unclean! Unclean!' (364) – and, crucially, by the ineffaceable 'red scar' of the holy wafer which

Van Helsing imprints on her forehead. Like the dermographic patient, then, Mina's body bears the double bloody inscription of the vampire/ male physician. However, while Lucy's body is positioned as a medium between the men and Dracula – between men and menstruation or, specifically, between men and *death* – Mina's figures as a cipher for the (at least superficially) oppositional male powers in *Dracula*: in her hypnotic trances, she becomes an amanuensis for Dracula and, as such, is intellectually regulated as 'the muse who leads the men to the vampire' (Bronfen, *Over Her Dead Body* 319) If Lucy, as menstrual subject, is controlled and contained through *corporeal* expulsion, Mina is circumscribed into the realm of paternal law through the 'ventriloquisation' of her body – an act which allays the threat of her *intellectual* prowess. It is, therefore, through the metaphor of the vampire – or, rather, *vampirisation* – that the menstrual monsters of Stoker's novel are constructed and, finally, contained.[8]

The flight into womanhood

Pierrette Fleutiaux's *Histoire de la chauve-souris* (*Hi/story of the Bat*) offers a counter-model to the pathologised menstrual monster occupying Stoker's text. Like *Dracula*, *Histoire de la chauve-souris* engages with the iconographies of vampirism haunting the discursive history of menstruation; and, like Stoker's Lucy, the protagonist of Fleutiaux's novel is positioned as representing a 'psychological study' for the various versions of the 'medical gaze' that she encounters on her journey of self-discovery. However, if, as Mulvey-Roberts suggests, *Dracula* is an 'anti-menstrual text' (78) then *Histoire de la chauve-souris* can be positioned as a recuperation of the menstrual narrative. Importantly, and in contrast to the ventriloquised narratives of vampirism/menstruation examined thus far, *Histoire de la chauve-souris* is distinguished by the first-person narrative of its unnamed narrator. Thus, the dual resonances of hi/story are important here insofar as they highlight the slippage between the 'fictional' and the 'historical' in psychoanalytic constructions of the case history.[9] Moreover, its episodic structure, combined with the patterns of ingestion/expulsion governing the bat's feedings, evoke a narrative pulse aligned with the rhythms of the menstrual cycle.[10] In *Histoire de la chauve-souris*, then, Fleutiaux re-works (and literalises) the Gothic motif of flight by endowing her heroine with wings (in the form of the bat entangled in her hair) at the moment of menarche; at the same time, the topos of the 'wandering womb' is reconfigured as Fleutiaux's 'bat-girl' roams the labyrinthine Gothic spaces of the narrative in search of a suitable space in which to speak (of) her menstrual identity.

The *chauve-souris* heroine begins her flight into womanhood from the confines of an archetypal Gothic tower. With its glacial, clean and pristine surfaces, this asylum-like reconfiguration of the Gothic castle symbolically inscribes the security and stability of the heroine's 'clean and proper' prepubescent positionality. In contrast, the dark, dank, *visceral* spaces beneath/outside the castle terrify the *chauve-souris* heroine precisely because they represent an unknown territory: 'I never go in the cellars underneath the tower. ... Without doubt, there are other dangers under the ground' (Fleutiaux 22).[11] The boundary between the white and pristine spaces of the heroine's room and the 'soiled', visceral geographies beyond its confines is signified by the window that she has been prohibited from opening. As Simone de Beauvoir proposes, 'the little girl, not yet in puberty, carries no menace, she is under no taboo and has no sacred character. ... But on the day she can reproduce, woman becomes impure; and rigorous taboos surround the menstruating female' (180). By opening the window, Fleutiaux's Gothic heroine transgresses menstrual taboo and initiates the arrival of the bat which becomes entangled in her hair. Indeed, the arrival of the bat brings about a moment of abject body horror: 'Against the nape of my neck I feel this other flesh ... I feel this great beating of imprisoned wings and a horror without name penetrates me' (27). In other words, this confrontation with abjection produces what Kristeva describes as 'one of those violent, dark revolts of being, directed against a threat that seems to emanate from an exorbitant outside or inside, ejected beyond the scope of the possible, the tolerable, the thinkable. It lies there, quite close, but it cannot be assimilated' (1). The arrival of the bat and the breaking of menstrual taboo literally shake the foundations of this Gothic castle as a codification of paternal law. As such, the menstrual heroine is thrust towards the threshold, 'the gaping hole of the window' (26) – a figuration of the grotesque 'gaping wound' upon which sexual difference is predicated.

Actualising Lucy's hallucinatory vision of a 'big bat ... buffeting its wings against the window' (185) in *Dracula*, the manifestation/infestation of the bat in *Histoire de la chauve-souris* recalls pre-industrial mythologies linking menarche to the bite of a vampire bat (see Lévi-Strauss 382), and makes explicit the link between vampirism and menstruation in Fleutiaux's text. However, while in Stoker's text the figure of the vampire serves as a metaphor for menstruation and its 'morbidification', the location of the bat in Fleutiaux's text marks a shift from a metaphoric to a metonymic inscription of menstrual identity. Although metaphor as a trope of verisimilitude ostensibly allows for a fluidity of meaning (via its dynamic of association), its economy of resemblance and substitution is circumscribed by a binary logic of figuration. As Domna C. Stanton highlights, the subversive limitations of metaphor are marked by its endorsement of a system of being as hidden 'essence': the ontological function of metaphor thus maintains rather than trans-forms bodies, identities and boundaries (161). In Fleutiaux's re-visioning, however, metonymy is a more useful trope in terms of re-thinking female psychic trans-formation precisely because it allows a mobile evo-cation of a different kind of being and contiguity across the mind/body split. By positioning the furry, membranous bat by her heroine's head (the traditional seat of rationalism), Fleutiaux's metonymical reconfigu-ration of menstrual subjectivity explodes the mind/body dualism over which nineteenth-century discourses of hysteria are navigated and which are signified most powerfully by the image of the severed head in *Dracula*.

This shift from a metaphorical to a metonymical configuration of menstrual identity is played out most powerfully in two key scenes in the early part of *Histoire de la chauve-souris*, both of which can be read as re-workings of the episodes from *Dracula* already explored. The first of these is the *chauve-souris* heroine's coming out at the annual village fes-tival, for which she is provided with a Gothic ball gown. With its 'widely open-necked collar, light and membranous' (39), the Medusa-like dress exteriorises the *chauve-souris* heroine's menstrual body: 'The fabric repulses me, shivers of disgust run over my skin' (39). A metaphorical iteration of the bat, this Gothic costume renders the heroine's body a *readable* and, therefore, *containable* spectacle; like Charcot's female patients the 'vampirised' *chauve-souris* heroine enacts a *masquerade* of her 'hysterical' (menstrual) identity. The dynamics of containment represented by the metaphorisation of the heroine's body are, accord-ingly, foregrounded by the brutal crucifixion/'staking' of the 'real' bat that descends upon the carnival festivities. Sacrifice, Kristeva proposes, 'merely extends the logic of taboo when the latter is perturbed' (95). Thus, in spite of the associations of the village festival with Bakhtinian notions of the carnival as the oldest topos of regenerative fiction, this is not a moment where objects are turned 'inside out' to expose the fragility of the dominant economy (Bakhtin 4), but a moment that functions to regulate and 'exorcise' difference by fixing bodies and iden-tities. This is foregrounded by the sinister enactment of the etymology of 'carnival' in the Latin 'to remove flesh': the flesh that is symbolically removed here is that of the heroine.

If the *chauve-souris* heroine's encounter at the festival reconfigures the vampiric/vampirised Lucy's confrontation with (and obliteration by) the Crew of Light in *Dracula*, then her next encounter is with the male vampire, incarnated here as *l'homme nu* (the naked man). Feeding off the discarded waste of a rubbish heap built on an abyss, *l'homme nu* embodies an animalistic, 'filthy' and predatory sexuality, signified by the creature living in his groin: 'In the hairs of his genitals, a strange, black creature moves, *similar to mine, and yet different*' (59; emphasis added). Echoing Count Dracula's vampirisation of Mina (with its inferences of enforced fellatio), *l'homme nu* pushes the *chauve-souris* heroine's head to the creature in his groin: 'suddenly, as I am still kneeling down, he grabs hold of my neck and pins my head against his legs and holds me firmly against him. ... Then, from the depths of this horror, I feel something moving softly against my head' (60). Nevertheless, Fleutiaux's reworking of this scene of violation problematises the subversion of bodily boundaries in Stoker's *Dracula*. In this metonymic reconfiguration of the Gothic heroine's menstrual identity, the positioning of the bat by the heroine's head foregrounds her encounter with *l'homme nu* as a moment of bodily and *psychic* violation. The conflicting positioning of the *chauve-souris* heroine as simultaneously vampiric and vampirised suggests a different kind of maidenhead through Fleutiaux's foregrounding of a psychic, rather than a solely physical, Gothic initiation into womanhood.

'Riveted between the Medusa and the abyss'[12]

Leaving behind the abyss, the *chauve-souris* heroine follows 'the murmur' of the bats and journeys towards an archaic place of verisimilitude in the dark, underground grotto-esque space of a bat cave. Mary Russo has detailed the confluence of the corporeal and the spatial in traditional figurations of the grotesque:

> As bodily metaphor, the grotesque cave tends to look like (and in the most gross metaphorical sense be identified with) the cavernous anatomical female body. ... Blood, tears, vomit, excrement – all the detritus of the body that is separated out and placed with terror and revulsion ... on the side of the feminine – are down there in that cave of abjection. (1–2)

However, although this grotto-esque space superficially offers a space in which the *chauve-souris* heroine can explore her metonymic relation to her bat/vampiric identity (in her attempts to hang upside down alongside the bats), the 'frozen' interior of the cave reiterates the glacial physical contexts of the initiatory Gothic tower. The extent to which this 'cavernous' space is a place of patriarchal colonisation, and the *chauve-souris* heroine's aspect as a vampire a result of ventriloquism, is foregrounded by the invasion of the speleologue, the first of various incarnations of the scientist/analyst who attempt to categorise and circumscribe both her body and her narrative. Metaphorically assuming the position of a gynaecologist, the speleologue catalogues the bats, turning his scrutiny to the *chauve-souris* heroine herself: 'He lifts his lamp directly above my face and looks at me' (60). As Carol A. Mossman argues, 'the first space to be penetrated, infiltrated, annexed, probed, and occupied is that of the maternal body. Her continent *must* remain steeped in darkness if the missionary light of civilization is to be cast upon it' (227; emphasis in original). Not only does the speleologue (as a reconfiguration of the Crew of Light in *Dracula*) attempt to contain and control the 'bat-girl' in his exploration of this 'dark continent' space but, after pulling her out of the cave and turning her up 'the right way', he returns her to another Gothic castle – another space of supervision and surveillance.

The extent to which the Gothic castles in *Histoire de la chauve-souris* codify the structures/strictures of paternal law and knowledge is exemplified by the *chauve-souris* heroine's confrontation with *le docteur introspecteur*. An absurd parody of the diagnostic positions taken up by Freud and Charcot, the doctor's specious questioning is a clichéd rendering of the psychoanalytic scene:

> 'Did you love your mother? ... And your father, a violent man no doubt? ... try hard to remember, has there not been an event in your childhood that marked you, an event that terrified you, leaving you with a lasting guilt complex?' ... He keeps talking, and other elements come and take their place *one by one in the structure*. (163–4; emphasis added)

In other words, the doctor attempts to (en)close her narrative within a hermeneutic derived from the 'structures' of the initial Gothic castle.[13] However, the *chauve-souris* heroine reclaims narrative agency by deflecting the analyst's inquisition. By requesting that the analyst touch her bat, she not only proves that it is 'real' but, at the very moment it vomits mucous onto his skin, disrupts the fragile borders of the psychoanalytic scene. The otherwise 'vampirised' heroine thus redeploys her 'vampiric' traits to destabilise the corporeal and discursive delineation of (her) subjectivity. It is, however, in her confrontation with the cineaste, who attempts to finally contain/circumscribe the 'enigma' of her menstrual identity, that the *chauve-souris* heroine ultimately returns and deflects the (medical) gaze. Crucially, the cineaste wants to exchange the furry, membranous (feminised) bat for 'a beautiful white gull' (169). In short, he wants to sanitise her narrative – to cleanse her corporeal *and* textual body. The stability of the narrative scene is again disrupted by the bat's secretions, this time supplemented by the abject discharge of the heroine's laughter: 'I hear myself laugh ... the inner emptiness on the verge of which I was standing frozen in a painful cramp *crumbles into dust* ... I laugh again, spewing out with each hiccup sweat, fear, jealousy, frenzy, I laugh, I laugh' (171; emphasis added). The image of dust here is, of course, an inversion of the Count's fate – and the metaphorical effacement of the menstrual subject – in *Dracula*.

Resisting vampirisation, then, the *chauve-souris* re-claims her bat/ vampiric identity as the source of her narrative agency. As she slowly unravels the hairs encasing her bat, it symbolically emerges from the head of the Medusa: 'You only have to look at the Medusa straight on to see her. And she's not deadly. She's beautiful and she's laughing' (Cixous 342). This re-positioning of the heroine's vampiric identity via her alignment with the Medusa corresponds to Gina Wisker's articulation of the radical potential of feminist re-visionings of the female vampire, where the 'transgression of gender boundaries, life/death, day/night behaviour ... is no longer abject, rejected with disgust to ensure identity', but 'enables us to recognise that the *Other is part of ourselves*' (168; emphasis in original). Fleutiaux's recuperation of the menstrual subject as vampire (bat) thus returns to pre-industrial mythologies to subvert the narratives of ventriloquism/vampirisation performed by Charcot's and Stoker's female vampires, and reinstates a more archaic form of ventriloquism – of 'speaking from the belly'[14] – that enables the menstrual subject to re-emerge as an agent in her own (hi)story.

Notes

1. While the distinction between the 'male Gothic' and the 'female Gothic' should be treated with some caution, it is helpful in thinking through Fleutiaux's self-positioning in relationship to the Gothic genre. For more on this see my 'Re-visioning the Gothic' (9–10).
2. As Jane Gallop proposes, the blood that flows in 'phallic fantasy' signifies 'defloration, wound as proof of penetration, breaking and entering, property damage. ... Menstrual blood is not a wound in the closure of the body; the menstrual flow ignores the distinction virgin/deflowered. ... In sadistic science there is no place for menstrual blood, for the latter marks woman as woman (virgin or not) with no need of man's tools' (83).

3. In Baudelaire's *Les Fleurs du mal* (1857 and 1861), the figure of the female vamp(ire) as prostitute recurs as a symbol of female predation – 'the perfect image of the savagery that lurks in the midst of civilisation' ('Painter' 36). 'Les Métamorphoses du vampire,' for example, portrays a sexually licentious vamp who, having sucked the 'marrow' from the poet's bones, undergoes a grotesque, abject metamorphosis before dematerialising into a pile of wizened bones.
4. 'Essentially different from "uncanniness", more violent, too, abjection is elaborated through a failure to recognize its kin; nothing is familiar, not even the shadow of a memory' (Kristeva 5).
5. In 'The Taboo of Virginity' (1918), Sigmund Freud describes how '[m]enstruation, especially its first appearance, is interpreted as the bite of some spirit-animal, perhaps as a sign of sexual intercourse with this spirit' (269).
6. Indeed, as Beizer notes, the most frequently used expression for menstruation in French texts is *être réglée*, meaning 'to be well ordered, steady, stable, fixed' (40).
7. The image of the Medusa is also linked to ancient cultural notions of menarche as brought on when the pubescent girl is bitten by the snake-goddess (see Creed 64).
8. Showalter identifies two key figurations of the female vampire in *Dracula* as encoding social and cultural anxieties about the New Woman of the 1880s and 1890s. Firstly, the portrayal of Lucy's 'sexual daring' identifies her with notions of the vampire as the 'nymphomaniac or oversexed wife who threatened her husband's life with her insatiable erotic demands'. Mina, on the other hand, represents the vampire as 'hysteric, the feminist intellectual whose sicknesses drain her family's energies' (180). The 'punishments' enacted against Lucy and Mina thus mirror their respective bodily and intellectual positionings.
9. Freud's most (in)famous case history, that of Dora (Ida Bauer), can be read as an archetypal Gothic narrative in its scripting of an adolescent girl on the brink of womanhood as an object of exchange between men. Here Freud explicitly positions Dora as a Gothic heroine in his reading of her first dream – 'the intention might have been consciously expressed in some such words as these: "I must fly from this house, for I see that my virginity is threatened here" ' ('Fragment' 123).
10. The heroine's compulsive preoccupation with feeding the bat also suggests that she is re-pathologised via an association with bulimia. For more on this see my 'Re-visioning the Gothic' (91–2).
11. All translations from this text are my own.
12. Cixous 341.
13. This metaphorical equation between topographical and discursive structures is more striking in the original French: 'Il parle encore, et d'autres éléments viennent prendre leur place *tour à tour dans l'édifice*.' (emphasis added).
14. The etymology of 'ventriloquism' is in the Latin *venter* (abdomen) and *loqui* (to speak).

Works cited

Auerbach, Nina. *Our Vampires, Ourselves*. Chicago: Chicago University Press, 1995.
Bakhtin, Mikhail. *Rabelais and his World*. Trans. Hélène Iswolsky. Bloomington: Indiana University Press, 1984.

Baudelaire, Charles. *Les Fleurs du Mal*. Intro. Claude Pichois. 2nd ed. Paris: Gallimard, 1996.

———. 'The Painter of Modern Life'. *The Painter of Modern Life and Other Essays*. Trans. and ed. Jonathan Mayne. 2nd ed. London: Phaidon, 1995. 1–41.

de Beauvoir, Simone. *The Second Sex*. Trans. and ed. H.M. Parshley. London: Pan–Picador, 1988.

Beizer, Janet. *Ventriloquized Bodies: Narratives of Hysteria in Nineteenth-Century France*. Ithaca: Cornell University Press, 1993.

Botting, Fred. *Gothic*. London: Routledge, 1996.

Bronfen, Elisabeth. *The Knotted Subject: Hysteria and its Discontents*. Princeton: Princeton University Press, 1998.

———. *Over Her Dead Body*. Manchester: Manchester University Press, 1992.

Cixous, Hélène. 'The Laugh of the Medusa'. Trans. Keith Cohen and Paula Cohen. *Feminisms: An Anthology of Literary Theory and Criticism*. Ed. Robyn R. Warhol and Diane Price Herndl. New Brunswick: Rutgers University Press, 1993. 334–49.

Creed, Barbara. *The Monstrous Feminine: Film, Feminism, Psychoanalysis*. London: Routledge, 1993.

Fleenor, Juliann E. Introduction. *The Female Gothic*. Ed. Juliann E. Fleenor. Montréal: Eden Press, 1983. 3–28.

Fleutiaux, Pierrette. *Histoire de la chauve-souris*. Paris: Julliard, 1975.

Foucault, Michel. *The Will To Knowledge: The History of Sexuality, Volume One*. Trans. Robert Hurley. London: Penguin, 1998.

Freud, Sigmund. 'Fragment of an Analysis of a Case of Hysteria ("Dora").' *Case Histories I: 'Dora' and 'Little Hans'*. Vol. 8 of *The Penguin Freud Library*. Ed. Angela Richards. Trans. Alix and James Strachey. 15 vols. London: Penguin, 1990. 31–164.

———. 'The Taboo of Virginity'. *On Sexuality*. Vol. 7 of *The Penguin Freud Library*. Ed. Angela Richards. Trans. Alix and James Strachey. 15 vols. London: Penguin, 1977. 261–83.

Gallop, Jane. *Feminism and Psychoanalysis: The Daughter's Seduction*. Basingstoke: Macmillan, 1982.

Grosz, Elizabeth. 'The Body of Signification'. *Abjection, Melancholia and Love: The Work of Julia Kristeva*. Ed. John Fletcher and Andrew Benjamin. London: Routledge, 1990. 80–103.

Kristeva, Julia. *Powers of Horror: An Essay on Abjection*. Trans. Leon S. Roudiez. New York: Columbia University Press, 1982.

Lévi-Strauss, Claude. *From Honey to Ashes: Introduction to a Science of Mythology, 2*. Trans. John and Doreen Weightman. London: Jonathan Cape, 1977.

Mossman, Carol A. *Politics and Narratives of Birth: Gynocolonization from Rousseau to Zola*. Cambridge: Cambridge University Press, 1993.

Mulvey-Roberts, Marie. '*Dracula* and the Doctors: Bad Blood, Menstrual Taboo and the New Woman'. *Bram Stoker: History, Psychoanalysis and the Gothic*. Ed. William Hughes and Andrew Smith. Basingstoke: Macmillan, 1998. 78–95.

Munford, Rebecca. 'Re-visioning the Gothic: A Comparative Reading of Angela Carter and Pierrette Fleutiaux'. PhD Diss. University of Exeter, 2003.

Russo, Mary. *The Female Grotesque: Risk, Excess and Modernity*. New York: Routledge, 1994.

Sade, Marquis de. *Justine, ou les malheurs de la vertu*. Intro. Gilbert Lely. Paris: Union Générale d'Éditions, 1969.

Showalter, Elaine. *Sexual Anarchy: Gender and Culture at the Fin de Siècle*. London: Virago, 1992.

Smith-Rosenberg, Carroll. *Disorderly Conduct: Visions of Gender in Victorian America*. Oxford: Oxford University Press, 1986.

Stanton, Domna C. 'Difference on Trial: A Critique of the Maternal Metaphor in Cixous, Irigaray, and Kristeva'. *The Poetics of Gender*. Ed. Nancy K. Miller. New York: Columbia University Press, 1986. 157–82.

Stoker, Bram. *Dracula*. Ed. and intro. Maurice Hindle. London: Penguin, 1993.

Wisker, Gina. 'Love Bites: Contemporary Women's Vampire Fictions'. *A Companion to the Gothic*. Ed. David Punter. Oxford: Blackwell, 2000. 167–79.

Part III
Appendix

20
A Guide to Bibliographical and Archival Resources for the Study of the Cultural History of Menstruation

Andrew Shail

As this history has been necessarily episodic and by no means exclusive, much remains to be added, and this chapter serves as a general guide for scholars engaging with aspects of the history of menstruation as part of cultural or medical history. Although we have not been able to include more than a nod to the wealth of anthropological work on menstruation in this collection, anthropological studies of menstrual beliefs are not discontinuous with studies of the cultural construction of menstruation. Anthropologists who have published specifically on, or are currently working on, menstruation include Alma Gottlieb (University of Illinois Urbana-Champaign), Judy Grahn (New College of California), Janet Hoskins (University of Southern California), Chris Knight (University of East London), Camilla Power (University of East London) and Dawn Starin (University College, London). This guide is split into two sections: a chronological discussion of major primary and archival sources followed by a list of secondary works (also see the lists of works cited in the main chapters).

Primary sources

Monica Green's *Women's Healthcare in the Medieval West: Texts and Contexts* (2000) contains a list of over 150 medieval gynaecological texts, and any scholar of medieval medicine will benefit from her translation of the medieval medical text *The Trotula*. Galenic medicine dominated in Europe between the twelfth and seventeenth centuries, and his *On the Usefulness of the Parts, On the Natural Faculties* and *On Semen* include discussions of menstruation. The latter is Galen's extensive debunking of Aristotle's theory that semen is the 'spirit' and menstrual blood the 'matter' in reproduction. These, along with other texts by Galen, Avicenna, and Hippocrates, continued to constitute much of early Modern medicine's understanding of the body and especially its reproductive functions, although major challenges were mounted to Galenism from around the early sixteenth century. While very few early Modern texts concentrate solely on the topic of menstruation, a wide variety of more general medical texts do contain important discussions. Two of the most important Early Modern texts which deal with menstruation are Isreal Spachius's *Gynaeciorum sive de Mulierum tum Vommunibus* (1597) and Martin Schurig's *Parthenologia Historico-Medica* (1729). Jean Astruc provides an eighteenth-century catalogue of gynaecological and obstetrical texts in volume four of his *Traité des Maladies des Femmes* (1765). See Cathy McClive's 'Bleeding Flowers and Waning Moons' (2004) for an analysis of the medical texts circulating in early Modern France and a comparison between the texts in Astruc's catalogue and a wider range of vernacular and Latin texts written by surgeons, anatomists, midwives and lay men and women (22–43; app. 1) and for information on medical these concerning menstruation defended at the Parisian Faculty between 1554 and 1733 (app. 2; app. 3). Peter Biller's 'A "Scientific" View of Jews from Paris around 1300' (2001) includes transcriptions and translations of the original texts that contributed to the formulation of the myth of Jewish male menstruation.

The nineteenth century produced an array of debates on menstruation as its medical parameters were fought over, and this is apparent from the multitude of medical reference books produced. Theses in medicine can often provide very useful contemporary surveys of medical thought; similarly, medical journals proliferated in the nineteenth century, and provide a map of ongoing discussions of menstruation. The medical student textbook of the nineteenth century has also preserved that century's proliferation of medical discourse in easily accessible form. Thomas Laycock's *Treatise on the Nervous Diseases of Women* (1840) is an excellent source for information on the mid-nineteenth century's first movements towards understanding menstruation as a total body cycle, as well as providing a detailed integration of menstruation into a total body schema. The works of Edward John Tilt between the 1850s and the 1880s also chart the development of medical thought on menstruation in the late nineteenth century, and

John Chapman's *Functional Diseases of Women* (1863) contains extensive examples of cases of treatments for dysmenorrhoea, leucorrhoea, vicarious menstruation and menorrhagia. The subject of discussion is often 'catamenia' rather than 'menstruation', or 'the climacteric' or 'change of life' rather than 'the menopause' – or much that is covered under the heading 'Diseases of Women'.

Notable medical history archives, particularly for nineteenth-century medicine, include the Barnes Medical Collection at the University of Birmingham, the Guy's Hospital Historical Medical Collections at King's College, London, the Medical Collection at Newcastle University Library, the Library of the Royal College of Physicians and Surgeons of Glasgow, and the Winterbottom Collection at Durham University. The richest archival resource is the Wellcome Library for the History and Understanding of Medicine (http://library.wellcome.ac.uk). The home of around 600,000 items pertaining to medical history, it is the best single place to start any work on the history of menstruation in the British Isles. German and French language texts are also in abundance. Works on subjects from the seventeenth century to the present day and from the effects of temperature on menstrual flow to the effects of menstruation on exam performance are to be found in the general collection, along with a wealth of texts on menstrual disorders, disturbances and diseases, and a good record of menstrual advice. Any historian of medical texts from as late as the eighteenth century may also need to be able to translate from Latin. In addition, a number of their Special Collections contain material relating to menstruation and are listed in their online sources leaflet 'Women's Health' (http://library.wellcome. ac.uk/ doc_WTL039963.html). The Library is home to some extremely rare historical works dating from as early as the mid-fifteenth century. The Library's collection of domestic medicine and 'receipt' or 'recipe' books from, in English, as early as the seventeenth century also include a range of 'emmenagogic' or 'emnagogic' preparations (having the power to bring on the incidence of menstruation). The midwifery text or 'System of Midwifery' also regularly discusses menstruation, and menstruation is a staple of general gynaecological works. Particularly relevant to the study of menstruation in the nineteenth century are the general practice case notes of James Mann Williamson on changing medicalisations of menstruation in the 1890s. With reference to the twentieth century, the papers of the Medical Women's Federation, containing material on all aspects of women's health from the 1900s to the 1990s, the Elizabeth Garrett Anderson papers, many of the papers of Frederick Parkes Weber on women's health between the 1890s and 1950s, and the findings of

R A McCance and E M Widdowson's 'Study of physical and emotional periodicity in women' (1929–30) will be of use. A determined search through the mass of correspondence received by Marie Stopes and Grantly Dick Read in the first half of the twentieth century will yield a great deal of detailed material. R J Hetherington's work between the 1960s and the 1980s on the Pill also contains papers on menstruation. Particularly because of the stark contrast between the two contemporary menstrual languages of 'getting it' and that behind menstrual suppression, also of significance is the commercial teenage menstrual advice booklet, many samples of which are held at the Wellcome Library and the Women's Library at London Metropolitan University.

Harry Finley's Museum of Menstruation and Women's Health, based in Maryland in the US, can be viewed by appointment only (MUM, PO Box 2398, Landover Hills Branch, Hyattsville, Maryland 20784–2398, USA; http://www.mum.org.) For the menstrual apparatus and the 'ephemera' (advertising, advice booklets, product designs, etc.) of the sanitary age the museum is unrivalled. It constitutes the best material culture collection on menstruation in the world, and many of the items can be viewed in the online version of the museum. The museum also contains a wealth of links to other archival sources in the US. The Material Culture and Advertising, Marketing and Commercial Imagery Collections at the Smithsonian Museum of American History also contain important information and artefacts on the history of sanitary products and advertising in the US. The Frank and Lillian Gilbreth Papers in Special Collections at Purdue University, including the report for Johnson & Johnson, are crucial evidence not just of late-1920s attitudes towards menstruation in the US but of the institutions and practices it was now seen as appropriate to bring to menstruation. The Ad*Access database of advertisements from the J. Walter Thompson Company Competitive Advertisements Collection of the John W. Hartman Center for Sales, Advertising, and Marketing History in Duke University's Rare Book, Manuscript, and Special Collections Library, online at Duke University (http://scriptorium.lib.duke.edu/adaccess), provides a great sample of menstrual advertising.

Recent medical works bear examination, both because the 'purpose' of menstruation is, being a function merely of contemporary medical discourses, not yet clear, and because 'egalitarian' gender beliefs are not the sole influence on medical research; such texts include Katharina Dalton's *Once a Month* (1978) and Allen Lein's *The Cycling Female* (1979). The articles of *Village Voice* journalist Karen Houppert, culminating in her recent taboo-breaker *The Curse* (1999), also bear inspection as examples of the contemporary menstrual preoccupations brought to bear on the taboo. Current beliefs about menstruation are also being charted and shaped by the Society for Menstrual Cycle Research at the University of Illinois in Chicago (http://www.pop.psu.edu/smcr).

Works cited

Astruc, Jean. *Traité des Maladies des Femmes*. Lyon, 1765.

Biller, Peter. 'A "Scientific" View of Jews from Paris around 1300'. *Micrologus* 9 (2001): 137–68.

Chapman, John. *Functional Diseases of Women. Cases Illustrative of a New Method of Treating Them Through the Agency of the Nervous System by Cold and Heat*. London: Trübner and Co., 1863.

Dalton, Katharina. *Once a Month: The Menstrual Syndrome, its Causes and Consequences*. Hassocks: Harvester, 1979.

Galen, Claudius. *On the Natural Faculties*. Trans. Arthur John Brock. London: Heinemann, 1916.

——. *On Semen*. Trans. Phillip De Lacy. Berlin: Akademie Verlag, 1992.

——. *On the Usefulness of the Parts*. Trans. Margaret Tallmadge May. 2 vols. New York: Classics of Medicine Library, 1996.

Green, Monica H. *Women's Healthcare in the Medieval West: Texts and Contexts*. Aldershot: Ashgate, 2000.

Houppert, Karen. *The Curse: Confronting the Last Unmentionable Taboo: Menstruation*. New York: Farrar, Strauss and Giroux, 1999.

Lein, Allen. *The Cycling Female: Her Menstrual Rhythm*. San Francisco: Freeman, 1979.

Laycock, Thomas. *A Treatise on the Nervous Diseases of Women; Comprising an Inquiry into the Nature, Causes and Treatment of Spinal and Hysterical Disorders*. London: Longman, Orme, Brown, Green and Longmans, 1840.

McClive, Cathy. 'Bleeding Flowers and Waning Moons: A History of Menstruation in France, c. 1495–1761'. PhD Diss. University of Warwick, 2004.

Schurig, Martin. *Parthenologia Historico-Medica*. Dresden and Leipzig, 1729.

Spachius, Israel, ed. *Gynaeciorum sive de Mulierum tum Communibus, tum Gravidarum, Parientium, et Puerperarum Affectibus et Morbis Libri*. Strasbourg, 1597.

The Trotula: A Medieval Compendium of Women's Medicine. Ed and trans. Monica H. Green. Philadelphia: Pennsylvania University Press, 2001.

Secondary sources

The following bibliography of works on the social and cultural history of menstruation lists the major works to date, so sometimes verging into fashion and design history, medical history, feminist theory, visual culture, or nursing. Work on the menstrual beliefs and practices of the twentieth century is dominated by the study of US contexts by US scholars, reflecting the part of the US in generating what is often referred to as the new menstrual language of mass-produced disposable sanitary paraphernalia, but also reflecting the ongoing importance of feminist scholarship in a country where menstrual taboo remains strong despite the ideological establishment of menstrual 'freedom'. For several reasons, including the availability and accessibility of sources and the guiding priority of mounting a challenge to the ongoing taboos of contemporary menstrual discourse, much of the published work deals with the twentieth century. Of these, Janice Delaney, Mary Jane Lupton and Emily Toth's *The Curse* (1976) is a ground-breaking, partially cross-cultural anthropological, second-wave feminist study of menstruation, required reading for any researcher although much of its statements have now been endlessly recycled. While the collection has few predecessors, in seeking to chart the invention of menstruation its immediate neighbour is Etienne van de Walle and Elisha P. Renne's *Regulating Menstruation* (2003), which provides an excellent analysis of medical and lay interventions in menstruation. Joan Jacobs Brumberg's ' "Something Happens to Girls" ' is both a history of the classed origins of the disposable sanitary towel and a good example of close observation of changes in the incidence of menstruation itself. In *The Disease of Virgins* (2004), Helen King traces the Hippocratic notion of 'greensickness' in antiquity through its re-introduction into Western Europe in the sixteenth century until the present day in order to demonstrate the importance of understanding the cultural meaning attributed to the absence of menstruation as a means to fully appreciate the resonance of the presence or functioning of menstrual bleeding. An analysis of the 'one-sex'/'two-sex' debate should start with Thomas Laqueur's *Making Sex* (1990) and Gianna Pomata's 'Menstruating Men' (2001). A chapter on Kotex in Susan Strasser's *Waste and Want* (1999) also lays vital ground-work for a study of menstruation in the disposable age.

al-Khalidi, A. 'Emergent Technologies in Menstrual Paraphernalia in Mid-Nineteenth-Century Britain'. *Journal of Design History* 14 (2001): 257–73.

Beusterien, John L. 'Jewish Male Menstruation in Seventeenth-century Spain'. *Bulletin of the History of Medicine* 73.3 (autumn 1999): 447–56.

Blackledge, Catherine. *The Story of V: Opening Pandora's Box*. London: Weidenfeld & Nicolson, 2003.

Brumberg, Joan Jacobs. *The Body Project: An Intimate History of American Girls*. New York: Random House, 1997.

——. ' "Something Happens to Girls": Menarche and the Emergence of the Modern American Hygienic Imperative'. *Journal of the History of Sexuality* 4.1 (July 1993): 99–127.

Bullough, Vern and Martha Voght. 'Women, Menstruation, and Nineteenth-Century Medicine'. *Bulletin of the History of Medicine* 47 (1973): 66–82.

Bynum, Caroline Walker. *Holy Feast and Holy Fast: The Religious Significance of Food to Medieval Women*. Berkeley: California University Press, 1988.

Chapman, Ruth. 'Attitudes of the Medical Profession Towards Menstruation and Menstrual Hygiene from the 1800s to the mid-1900s'. BSc Diss. Wellcome Institute for the History of Medicine, 1991.

Crawford, Patricia. 'Attitudes to Menstruation in Seventeenth-Century England'. *Past and Present* 91 (May 1981): 47–73.

Dean-Jones, Lesley. *Women's Bodies in Classical Greek Science*. Oxford: Clarendon, 1994.

Delaney, Janice, Mary Jane Lupton and Emily Toth. *The Curse: A Cultural History of Menstruation*. New York: Mentor, 1976.

Ehrenreich, Barbara and Dierdre English. *For Her Own Good: 150 Years of the Experts' Advice to Women*. London: Pluto, 1979.

Farrel-Beck, Jane and Laura Klosterman Kidd. 'The Roles of Health Professionals in the Development and Dissemination of Women's Sanitary Products, 1880–1940'. *Journal of the History of Medicine and Allied Sciences* 51.3 (July 1996): 325–52.

Fausto-Sterling, Anne. *Myths of Gender: Biological Theories About Men and Women*. New York: Basic, 1985.

Freidenfelds, Lara. 'Materializing the Modern, Middle-Class Body: Menstruation in the Twentieth Century United States'. PhD Diss. Harvard, 2003.

Gottlieb, Alma. 'American Premenstrual Syndrome: A Mute Voice'. *Anthropology Today* 4.6 (1988): 10–13.

Gottlieb, Alma and Thomas Buckley, eds. *Blood Magic: The Anthropology of Menstruation*. Berkeley: California University Press, 1988.

Healy, Margaret. 'Dangerous Blood: Menstruation, Medicine and Myth in Early Modern England'. *National Healths: Gender, Sexuality and Health in a Cross-Cultural Context*. Ed. Michael Worton and Nana Wilson-Tagoe. London: UCLP, 2004. 86–94.

Johnson, Willis. 'The Myth of Jewish Male Menses'. *Journal of Medieval History* 24.3 (1998): 273–95.

Jones, I.H. 'Menstruation: The History of Sanitary Protection'. *Nursing Times* 76 (1980): 407–8.

Kidd, Laura Klosterman. 'Menstrual Technology in the United States, 1854 to 1921'. PhD Diss. Iowa State University, 1994.

King, Helen. *The Disease of Virgins: Green Sickness Chlorosis and the Problems of Puberty*. London: Routledge, 2004.

———. *Hippocrates' Women: Reading the Female Body in Ancient Greece*. London: Routledge, 1998.

Laqueur, Thomas. *Making Sex: Body and Gender from the Greeks to Freud*. Cambridge: Harvard University Press, 1990.

Lee, Janet and Jennifer Sasser-Coen. *Blood Stories: Menarche and the Politics of the Female Body in Contemporary US Society*. London: Routledge, 1996.

Lord, Alexandra. ' "The Great Arcana of the Deity": Menstruation and Menstrual Disorders in Eighteenth-Century British Medical Thought'. *Bulletin of the History of Medicine* 73.1 (Spring 1999): 38–63.

Lupton, Mary Jane. *Menstruation and Psychoanalysis*. Chicago: University of Illinois Press, 1993.

Martin, Emily. *The Woman in the Body: A Cultural Analysis of Reproduction*. Boston: Beacon, 1992.

Martin, Michelle. 'Postmodern Periods: Menstruation Media in the 1990s'. *The Lion and the Unicorn* 23.3 (1999): 294–5.

Martin, Michelle and Claudia Nelson, ed. *Sexual Pedagogies: Sex Education in Britain, Australia, and America, 1879–2000.* New York: Palgrave Macmillan, 2004.

McCracken, Peggy. *The Curse of Eve, the Wound of the Hero: Blood, Gender, and Medieval Literature.* Philadelphia: Pennsylvania University Press, 2003.

Merskin, Debra. 'Adolescence, Advertising and the Ideology of Menstruation'. *Sex Roles* 40.11 (June 1999): 941–57.

Pomata, Gianna. 'Menstruating Men: Similarity and Difference of the Sexes in Early Modern Medicine'. *Generation and Degeneration: Tropes of Reproduction in Literature and History from Antiquity to Early Modern Europe.* Ed. Valeria Finucci and Kevin Brownlee. Durham: Duke University Press, 2001. 109–52.

Showalter, Elaine. *The Female Malady: Women, Madness and Culture in England, 1830–1980.* New York: Pantheon, 1986.

Showalter, Elaine and English Showalter. 'Victorian Women and Menstruation'. *Victorian Studies* 14 (1970–71): 83–9.

Shuttle, Penelope and Peter Redgrove. *The Wise Wound: Menstruation and Everywoman.* 1978. Harmondsworth: Penguin, 1980.

Strange, Julie-Marie. 'The Assault on Ignorance: Teaching Menstrual Etiquette in England, c. 1920s to 1960s'. *Social History of Medicine* 14.2 (August 2001): 247–65.

——. 'Menstrual Fictions: Languages of Medicine and Menstruation, c. 1850–1930'. *Women's History Review* 9.3 (Spring 1996): 607–28.

Strasser, Susan. *Waste and Want: A Social History of Trash.* New York: Metropolitan Books, 1999.

Troyer, Kristin de, *et al.*, eds. *Wholly Woman, Holy Blood: A Feminist Critique of Purity and Impurity.* London: Trinity, 2003.

Valdisseri, R.O. 'Menstruation and Medical Theory: an Historical Overview'. *Journal of the American Medical Women's Association* 38.3 (1983): 66–70.

van de Walle, Etienne. 'Flowers and Fruits: Two Thousand Years of Menstrual Regulation'. *Journal of Interdisciplinary History* 28.2 (Autumn 1997): 183–203.

van de Walle, Etienne and Elisha P. Renne, eds. *Regulating Menstruation: Beliefs, Practices, Interpretations.* Chicago: Chicago University Press, 2003.

Verbrugge, Martha. 'Gym Periods and Monthly Periods: Concepts of Menstruation in American Physical Education 1900–1940'. *Body Talk: Rhetoric, Technology, Reproduction.* Ed. Mary M Lay *et al.* Madison: Wisconsin University Press, 2000. 67–97.

Wasserfall, Rahel R, ed. *Women and Water: Menstruation in Jewish Life and Law.* London: New England University Press, 1999.

Index

abdomen, 42, 43, 270n
abdominal masses, 39, 43, 44, 46, 48
abjection, 259–62, 267–9
abortifacients, 46, 53
abortion, 91, 105
    induction of, 14
    menstruation as mini abortion, 119
Académie Royale des Sciences, 85
accoucheurs (male midwives), 84
acupuncture, 39, 40
adolescence, 215–22, 264–7
advertising, 1–2, 4, 113, 233, 239, 244, 252, 255–6, 278
aetiology (causes and origins of disorder), 39, 41, 43–4, 45, 46, 47, 54, 107, 230–1
agape, 153
Albert the Great, 65
Alcoff, Linda, 3
Alcott, Louisa May, 216, 218
Alcott, William, *see Young Wife, The*
Alfasi, Isaac ben Jacob, 191–2
Allbutt, Henry Arthur, *see Wife's Handbook, The*
Allbutt, T. Clifford, 231
All-Inclusive Good Prescriptions for Women, 47
al-Majusi, 58
amenorrhoea, 20, 55–6, 107–8, 131
amenorrhoeal insanity, 108
American Eugenics Society, 137
American Home Products Corporation, 140
American Physiological Society, *see* Physiological Society
anaemia, 135–6, 231
    see also iron-deficiency anaemia
ananku, South Indian concept of, 176–86
anatomy, 3, 19, 53–7, 78, 83, 85, 93, 96, 118, 202, 219, 227, 228, 267
Anatomy of the Pig, 54
ancient Greece, 13–36

Anderson, Elizabeth Garrett, 111, 277
Andrieu, Pierre, 56
'Anorexics, Holy', 158
Anstie, Francis, 106
anus, 60, 80–1
    see also bleeding, from anus
Aphrodite, 162, 164–7, 173
aprons, rubber, 243, 253
Aquinas, Thomas, 159n
Archbishop Caesarious of Arles, 152
Arderne, John, 57
Aristotelian thought, 26, 35, 36n, 58, 68, 71, 72, 78,
Aristotle, 25–36, 57, 61, 66, 276
    as two-seed theorist, 28
    Generation of Animals, 29, 30
    Metaphysics, 33
    see also one-seed theory
artificial fertilisation, 121
Ashkenaz, 194
Astruc, Jean, *see Traité des Maladies des Femmes*
asylum, 102, 106–10, 265
Athena, 151

Babylonian Talmud, 152–3, 191, 195
Bachofen, Johann Jakob, 166
bacon factories, 117
bacteriology, 239
balance
    bodily, 14, 39, 47–9, 67
    cosmic, 67–8, 70
    mental/physical, 108
    see also imbalance, Blood, *and qi*
balancing, menstrual, 38–49
Balfour, Francis Maitland, 120
Bara, Theda, 225, 235, 238–9
Barnes, Robert, 105, 106, 277
bathing, ritual, 59, 163, 171, 181, 182, 189
    see also dressing, ritual, '*mikveh*', *and* seclusion, ritual
Bathory, Countess Elizabeth, 149

CPSIA information can be obtained at www.ICGtesting.com
Printed in the USA
LVOW10*1450100616

492098LV00011B/76/P